Fundamentals of
PLANT PHYSIOLOGY

THE AUTHORS

Mr. Amit Kumar Singh obtained his M.Sc. (Ag) Agronomy from Acharaya Narendra Deva University of Agriculture and Technology, Kumarganj, Ayodhya (U.P.). He has qualified ARS (ICAR-NET) conducted by Agricultural Scientists Recruitment Board, New Delhi. Currently working as Block Technical Manager (BTM), ATMA, Paliganj, Patna. He worked as Senior Research Fellow (SRF) at ICAR-Research Complex for Eastern Region, Patna in research project sanctioned by Bill Gates Foundation, United States. He has vast experience in popularization of Agricultural farming techniques in the farmer field. He has published two books and number of research papers in national and international journals of repute and also organized/ participated various training programmes, seminars, symposia and workshops.

Fundamentals of PLANT PHYSIOLOGY

— Author —

Amit Kumar Singh

ASTRAL

Publishing Authors Across the Globe

2025

Daya Publishing House®

A Division of

Astral International Pvt. Ltd.

New Delhi – 110 002

ISBN : 978-93-5919-780-7

Published by : **Daya Publishing House®**
A Division of
Astral International Pvt. Ltd.
– ISO 9001:2015 Certified Company –
4736/23, Ansari Road, Darya Ganj
New Delhi-110 002
Ph. 011-43549197, 23278134
E-mail: info@astralint.com
Website: www.astralint.com

Printed at : **Replika Press Pvt Ltd**

Preface

Plant physiology is a cornerstone of understanding the life processes of plants, and it holds profound significance in various scientific and practical fields. At its core, plant physiology explores how plants function, from the molecular level of cellular processes to the systemic interactions within the plant and its environment. Understanding these physiological mechanisms is crucial not only for advancing basic biological science but also for improving agricultural productivity, sustainable food security, and environmental conservation. As plants are the primary producers of oxygen and food on Earth, their physiological processes–such as photosynthesis, respiration, and nutrient uptake–are central to the global carbon and nutrient cycles. In agriculture, for example, knowledge of plant physiology underpins the development of better crop varieties, improved irrigation methods, and more efficient use of fertilizers. By understanding how plants respond to water stress, temperature fluctuations, and soil conditions, plant physiologists can contribute to designing more resilient crops, enhancing agricultural output, and addressing the challenges of climate change. Furthermore, plant physiology plays a critical role in advancing biotechnology and environmental sciences. With the growing need for sustainable practices in the face of global environmental concerns, plant physiology provides the foundation for innovations like genetic modification, bioengineering, and the development of crops with enhanced nutritional content or resistance to pests and diseases. In addition, plant physiological research is key to tackling broader environmental issues, such as soil degradation, deforestation, and the loss of biodiversity. Plants are integral to the health of ecosystems, and a deep understanding of their physiological responses to environmental stresses helps in the development of strategies to mitigate the effects of environmental pollution, habitat destruction, and climate change. Whether through enhancing crop resilience or optimizing the carbon sequestration potential of plants, the significance of plant physiology extends far beyond the laboratory, influencing the future of agriculture, biotechnology, and environmental sustainability.

Fundamentals of Plant Physiology provides an in-depth exploration of the core concepts and processes that govern the life and function of plants. With a comprehensive approach, this book covers essential topics such as water and mineral nutrition, photosynthesis, respiration, plant growth and development, and the intricate mechanisms behind plant transport and communication. The chapters also delve into contemporary areas like plant biotechnology, secondary metabolites, and the broader ecological and evolutionary perspectives that influence plant function. By integrating both theoretical principles and practical insights, the book offers readers a thorough understanding of plant biology that is vital for students, researchers, and professionals in the field. This book is designed for students studying, botany, biochemistry, agriculture, environmental science, and biotechnology. It is suitable for undergraduate and postgraduate courses in plant physiology and related fields, serving as both a text cum reference book. Additionally, this book would be immensely useful to researchers and professionals working in agriculture, horticulture, and plant biotechnology.

We would like to express our sincere gratitude to all the individuals, as well as the various books, manuals, periodicals, newsletters, magazines, journals, and other resources, that contributed to the preparation of this book. Despite our best efforts, we acknowledge that some errors may have occurred during the compilation and editing process. We welcome any further queries, constructive suggestions, and feedback aimed at improving the book, and we will be deeply appreciative of such contributions.

Contents

Chapter 1

Plant Physiology: An Introduction

Plant Physiology: An Overview

Plant physiology is a sub-discipline of botany that focuses on the study of the vital processes and functions that occur within plants. It explores how plants grow, develop, reproduce, and interact with their environment, all while maintaining the intricate balance necessary for survival. As autotrophic organisms, plants are central to life on Earth, forming the foundation of most ecosystems by producing the oxygen we breathe and the food we consume. They play a crucial role in the global carbon and water cycles, influencing climate regulation, soil health, and biodiversity. Understanding plant physiology not only helps to unravel the basic biological principles of life but is also vital for addressing global challenges such as food security, climate change, and sustainable agricultural practices. At the heart of plant physiology is the concept that plants are highly dynamic systems. They are constantly interacting with their environment–responding to light, gravity, water availability, and nutrient supply. Plants carry out essential biochemical processes such as photosynthesis, respiration, and nutrient uptake, which enable them to synthesize food, grow, and reproduce. These processes are tightly regulated by a suite of molecular signals, such as hormones and other chemical messengers, that coordinate cellular activities across the entire organism. For example, the hormone auxin governs plant growth and response to light, while gibberellins control seed germination and flowering. Moreover, plants must cope with environmental challenges like drought, extreme temperatures, herbivory, and disease. Plant physiology, therefore, encompasses not just the basic processes of life, but also the adaptive mechanisms that allow plants to survive in diverse habitats and climates. This includes complex mechanisms like transpiration, which aids in water transport and cooling, or the production of secondary metabolites that help plants defend against predators and pathogens. The field of plant physiology is inherently interdisciplinary, bridging the gaps between biochemistry, molecular biology, genetics, ecology, and environmental science. As we face a rapidly changing world, with rising populations, shifting climates, and increasing demand for natural resources, understanding how plants function at the physiological level is more important than ever. Research in plant physiology underpins innovations in crop breeding, sustainable agriculture, environmental conservation, and even biotechnology. By uncovering the molecular and biochemical bases of plant behavior, scientists can develop crops that are more resistant to diseases, pests, and environmental stresses, ultimately helping to ensure food security and ecosystem health for future generations. Thus, plant physiology is not merely the study of plants in isolation, but rather a lens through which we can understand the interconnectedness of life on Earth and the critical roles plants play in maintaining the delicate balance of our planet's ecosystems. Through continued exploration of plant processes, we unlock new avenues for improving agricultural productivity, restoring damaged environments, and managing natural resources more sustainably.

History and Scope of Plant Physiology

History of Plant Physiology

The history of plant physiology dates back to the ancient period when early philosophers and scientists began to seek explanations for how plants grow, reproduce, and interact with their environment. The history of plant physiology has evolved from simple observations and basic experimental studies to a sophisticated understanding of the molecular, biochemical, and ecological processes that regulate plant life. Today, plant physiology is a dynamic and interdisciplinary field that contributes to a wide range of scientific advancements, from agricultural innovation to ecological sustainability and biotechnology. As the world faces challenges such as climate change, food security, and resource conservation, plant physiology will continue to play a critical role in developing solutions that ensure the health and sustainability of our plant-based ecosystems.

While much of early plant science was speculative, several key discoveries paved the way for modern plant physiology.

Ancient Observations

- ☆ Early civilizations, such as the Egyptians and Greeks, were keen observers of nature, and they made rudimentary connections between plant growth and the environment. However, the understanding of plant life processes was often limited to practical knowledge, like the use of plants for food, medicine, and religious rituals.

- ✰ In the 4th century BCE, Aristotle and later Theophrastus, who is often considered the "father of botany," made the first attempts to systematically classify and describe plant structures, though they had no concept of physiological processes as we understand them today.

17th and 18th Centuries – The Beginnings of Experimental Plant Physiology

- ✰ The scientific revolution of the 17th century laid the groundwork for more rigorous experimental methods. Jan Baptista van Helmont, in the early 1600s, conducted famous experiments on plant growth. He concluded that the increase in plant weight was due to water absorption, showing a rudimentary understanding of how plants utilize water for growth.
- ✰ In the 18th century, Joseph Priestley discovered that plants release oxygen, an important observation that later contributed to the understanding of photosynthesis. Stephen Hales, often regarded as the first plant physiologist, is credited with the first systematic experimental studies on plant transpiration, water uptake, and the role of air in plant respiration.

19th Century – The Golden Age of Plant Physiology

- ✰ The 19th century saw rapid advancements in plant physiology as scientists began to unravel the complex processes of plant life. The discovery of photosynthesis is often considered one of the defining moments in the history of plant physiology.
- ✰ Jean Senebier (1779) and Nicolas-Théodore de Saussure (1804) provided important insights into the role of carbon dioxide and sunlight in the process of photosynthesis.
- ✰ Fritz Went (1926) was a pioneer in the study of plant hormones (phytohormones) and how they control growth and development. His discovery of auxin, the first plant hormone, marked the beginning of understanding the chemical regulation of plant processes.
- ✰ Charles Darwin and his son Francis Darwin contributed significantly to our understanding of phototropism (the growth of plants towards light), further emphasizing the role of light in plant growth and development.

20th Century to Present – Molecular and Genetic Advances

- ✰ The 20th century witnessed the development of molecular biology, genetics, and biochemistry, leading to a much deeper understanding of the biochemical pathways underlying plant physiology. The discovery of DNA and the subsequent development of genetic engineering revolutionized plant science.
- ✰ The structure of chlorophyll and the biochemical pathway of photosynthesis were elucidated, with key contributions from scientists

like Melvin Calvin, who won the Nobel Prize in Chemistry in 1961 for his work on the Calvin cycle.

- The discovery of plant hormones, including gibberellins, cytokinins, ethylene, and abscisic acid, expanded our understanding of plant growth regulation.
- Modern plant physiology also benefits from technological advances such as genomic sequencing, CRISPR-Cas9 gene editing, and transgenic plants, which have enabled researchers to manipulate plant genetic material and study their physiological processes in unprecedented detail.

Scope of Plant Physiology

Plant physiology is a broad and interdisciplinary field that encompasses many aspects of plant life. It deals with both the fundamental mechanisms that control plant functions and their applications in agriculture, horticulture, and biotechnology. The scope of plant physiology can be categorized into several key areas:

- **Cellular and Molecular Physiology**: This area focuses on the biochemical and molecular mechanisms that govern cellular processes within plants. It includes the study of enzymes, metabolic pathways, protein synthesis, and genetic regulation of plant cells. Membrane transport, signal transduction, and photosynthesis are central topics, along with the role of plant hormones in regulating gene expression and physiological responses.
- **Growth and Development**: Plant growth is controlled by the coordination of cell division and expansion, and the plant's ability to sense and respond to environmental signals. This includes the study of tropisms (such as phototropism and gravitropism), germination, flowering, and fruit ripening. Plant development is influenced by genetic programming, environmental cues (such as light, temperature, and water availability), and the interaction of hormones like auxins, cytokinins, and gibberellins.
- **Photosynthesis and Respiration**: Photosynthesis is the process by which plants convert light energy into chemical energy, which is stored as sugars and used to fuel cellular activities. The molecular mechanisms of light absorption, electron transport, and carbon fixation are key areas of study. Cellular respiration, the process by which plants break down glucose to release energy in the form of ATP, is equally important for sustaining plant growth and metabolism.
- **Water and Nutrient Uptake**: Plants are highly dependent on the uptake of water and essential nutrients from the soil. The study of water transport (via the xylem), mineral uptake (via the roots), and the role of mycorrhizal fungi in nutrient acquisition are critical topics. Transpiration and its role in water movement, cooling, and nutrient transport are also important aspects of plant water physiology.

- ☆ **Stress Physiology**: Plants are constantly exposed to various environmental stresses, including drought, salinity, extreme temperatures, and pathogens. Studying how plants respond to these stresses is crucial for improving crop resilience and developing stress-tolerant plants. Abiotic stress responses may include adjustments in metabolic processes, protein synthesis, and hormone signaling, while biotic stress responses often involve the activation of defense mechanisms such as the production of secondary metabolites and the formation of physical barriers.
- ☆ **Ecophysiology**: This subfield focuses on the interactions between plants and their environment, including how plants adapt to different climates, soil types, and environmental conditions. It includes the study of plant-water relations, photosynthetic efficiency, and the impact of environmental factors such as light, temperature, and CO_2 levels on plant function.
- ☆ **Agricultural and Industrial Applications**: A major scope of plant physiology is its application in agriculture to improve crop yield, quality, and resistance to pests and diseases. Advances in plant physiology have led to the development of improved irrigation techniques, better crop varieties, and strategies for managing plant diseases. The biotechnology aspect of plant physiology focuses on genetic modification, genetic engineering, and the use of plants for biofuels, pharmaceuticals, and bioremediation.
- ☆ **Plant Reproduction**: The study of plant reproduction covers both sexual and asexual reproduction, including mechanisms like pollination, seed formation, and vegetative propagation. Understanding the physiological processes involved in seed dormancy, germination, and flowering is essential for optimizing crop production and breeding new plant varieties.

The Cell: The Basic Unit of Life in Plants

The cell is the fundamental structural and functional unit of all living organisms, including plants. In plants, as in all eukaryotic organisms, the cell is the building block that carries out the essential processes of life. The plant cell is the basic unit of life in plants, intricately designed to carry out a range of essential functions, from photosynthesis and growth to transport and defense. Each organelle within the cell plays a specific and vital role in the life of the plant. Understanding plant cell structure and function is fundamental to understanding plant biology and forms the foundation for fields such as agriculture, biotechnology, and ecology. Advances in plant cell research continue to offer new insights into improving crop productivity, developing stress-resistant plants, and understanding plant-environment interactions. Plant cells are highly specialized to perform specific functions that support the growth, reproduction, and survival of the plant. Understanding the structure and function of plant cells is crucial for comprehending how plants grow, respond to their environment, and carry out complex physiological processes such as photosynthesis, respiration, and nutrient uptake.

Structure of Plant Cells

Plant cells, like animal cells, are eukaryotic, meaning they have a well-defined nucleus and membrane-bound organelles. However, plant cells differ from animal cells in several key structural features, which are critical to their function. The major components of a plant cell include:

- **Cell Wall:** The cell wall is one of the most distinctive features of plant cells. It is a rigid, protective outer layer made primarily of cellulose, a polysaccharide, along with other materials such as hemicellulose and lignin. The cell wall provides structural support, determines cell shape, and protects the cell from mechanical stress and pathogens. It also plays a role in regulating the movement of substances into and out of the cell. In addition to the primary cell wall, many plant cells also have a secondary cell wall, which is thicker and often contains lignin, providing additional strength and rigidity, especially in woody plants.
- **Plasma Membrane:** Beneath the cell wall lies the plasma membrane, a semi-permeable lipid bilayer that surrounds the cytoplasm. The plasma membrane regulates the movement of water, ions, nutrients, and waste products into and out of the cell. It is involved in cell signaling, communication, and interactions with the surrounding environment.
- **Nucleus:** The nucleus is the control center of the plant cell, containing most of the plant's genetic material in the form of DNA. The nucleus is surrounded by a double membrane called the nuclear envelope, which has pores that allow selective exchange of materials between the nucleus and the cytoplasm. The nucleolus, located within the nucleus, is involved in the synthesis of ribosomal RNA.
- **Cytoplasm and Cytosol:** The cytoplasm is the gel-like substance that fills the interior of the cell, consisting of the cytosol (the liquid portion) and various organelles. The cytoskeleton within the cytoplasm provides structure, facilitates intracellular transport, and supports cell movement.
- **Vacuole:** The vacuole is a large, membrane-bound organelle found in plant cells that contains cell sap, a solution of water, salts, sugars, and other substances. It plays a key role in maintaining turgor pressure (the pressure of the cell's contents against the cell wall), which helps the plant maintain its shape and structural integrity. The vacuole also serves in storage (of water, sugars, pigments, and waste products) and detoxification (by isolating harmful substances).
- **Chloroplasts:** Chloroplasts are the sites of photosynthesis, the process by which plants convert light energy into chemical energy stored in glucose. Chloroplasts contain the green pigment chlorophyll, which captures light energy, and other pigments involved in the absorption of light. Chloroplasts have a double membrane structure and contain internal membrane systems called thylakoids, which are organized into stacks known as grana. These membranes house the light-dependent reactions

of photosynthesis. The stroma, the fluid-filled space surrounding the thylakoids, is where the Calvin cycle (the light-independent reactions) occurs, converting carbon dioxide into glucose.

- ☆ **Mitochondria:** Mitochondria are known as the "powerhouses" of the cell because they are the sites of cellular respiration, where glucose is broken down into ATP (adenosine triphosphate), the primary energy currency of the cell. Mitochondria are membrane-bound organelles with an inner membrane folded into cristae to increase surface area for energy production.
- ☆ **Endoplasmic Reticulum (ER):** The endoplasmic reticulum (ER) is a network of membranes involved in protein and lipid synthesis. There are two types of ER: Rough ER, studded with ribosomes, is involved in protein synthesis and modification. Smooth ER lacks ribosomes and is involved in lipid synthesis, detoxification, and calcium storage.
- ☆ **Golgi Apparatus:** The Golgi apparatus is responsible for modifying, sorting, and packaging proteins and lipids synthesized in the ER. It processes these molecules and sends them to their final destinations within or outside the cell.
- ☆ **Ribosomes:** Ribosomes are the sites of protein synthesis. They can either be attached to the rough ER or free-floating in the cytoplasm. Ribosomes read messenger RNA (mRNA) and assemble amino acids into proteins according to genetic instructions.
- ☆ **Peroxisomes:** Peroxisomes are involved in the breakdown of fatty acids, detoxification of harmful substances, and the conversion of hydrogen peroxide (a byproduct of cellular metabolism) into water and oxygen.

Function of Plant Cells

Each component of the plant cell plays a specific role that contributes to the overall function of the plant. Some key physiological functions of plant cells include:

- ☆ **Photosynthesis:** The primary function of plant cells, especially those containing chloroplasts, is photosynthesis. This process allows plants to convert sunlight, carbon dioxide, and water into glucose (a form of sugar) and oxygen. Photosynthesis occurs in the chloroplasts, where light energy is captured by chlorophyll and used to drive the reactions that convert carbon dioxide and water into glucose and oxygen.
- ☆ **Growth and Development:** Plant cells divide and enlarge to facilitate the growth of the plant. The growth of plant cells is largely driven by turgor pressure (maintained by the vacuole) and the cell wall, which restricts expansion but allows for controlled growth. As cells mature, they differentiate into specialized types of cells, forming tissues such as xylem, phloem, and epidermis, each with specific functions in transport, protection, and support.

- ☆ **Transport of Water, Nutrients, and Sugars:** The xylem cells transport water and minerals from the roots to the rest of the plant, while the phloem cells transport the products of photosynthesis (primarily sugars) from the leaves to other parts of the plant. Transport processes are regulated by various mechanisms, including osmosis, active transport, and bulk flow through vascular tissues.
- ☆ **Storage:** Plant cells store various substances such as starch, lipids, and proteins. The vacuole plays a key role in storing water, sugars, and minerals, while plastids (like amyloplasts) store starch in specialized cells.
- ☆ **Defense:** Plant cells have mechanisms for responding to environmental stresses and defending against pathogens. The cell wall serves as a physical barrier, and cells can also produce secondary metabolites (like alkaloids, tannins, and terpenoids) that deter herbivores or inhibit the growth of pathogens.

Levels of Organization in Plants: Cells, Tissues and Organs

Plants, like all multicellular organisms, are organized into distinct structural and functional levels. These levels of organization–cells, tissues, and organs–work together to enable plants to perform essential life functions such as growth, reproduction, nutrient acquisition, and environmental adaptation. Understanding these levels provides insight into how plants maintain their form, carry out complex physiological processes, and interact with their environment.

The Cellular Level: The Basic Unit of Life

The **cell** is the most fundamental unit of life in all organisms, and in plants, it is no different. Plant cells are the building blocks of plant tissues and organs. They are highly specialized to carry out specific functions necessary for plant growth and survival.

Features of Plant Cells

- ☆ **Cell Wall**: The rigid outer layer that provides structural support and defines the shape of the cell.
- ☆ **Chloroplasts**: Organelles containing chlorophyll, responsible for photosynthesis.
- ☆ **Vacuole**: A large, fluid-filled organelle that helps maintain turgor pressure and stores nutrients, waste products, and other substances.
- ☆ **Nucleus**: The control center containing genetic material (DNA) that governs cell functions.
- ☆ **Cytoplasm**: The jelly-like substance where various metabolic activities occur and organelles are suspended.

Plant Cell Specialization

Each plant cell is specialized to perform specific tasks within the plant. For example:

- ☆ **Parenchyma Cells**: Involved in photosynthesis, storage, and wound healing.
- ☆ **Collenchyma Cells**: Provide structural support, especially in growing parts of the plant.
- ☆ **Sclerenchyma Cells**: Provide strength and rigidity in mature tissues.
- ☆ **Xylem and Phloem Cells**: Specialized for water and nutrient transport (xylem) and the movement of sugars (phloem).

These cells work in concert to form **tissues**, the next level of organization.

The Tissue Level- Groups of Specialized Cells

Plant tissues are groups of cells that work together to perform specific functions. There are two major categories of plant tissues: meristematic tissue and permanent tissue.

Meristematic Tissue (Growth Tissue)

Meristems are regions of active cell division, where new cells are continuously produced to facilitate plant growth. Meristematic tissue is undifferentiated (cells do not have a specific function yet) and is found in regions of the plant where growth occurs, such as in the tips of roots and shoots (apical meristems) or along the sides of stems in woody plants (lateral meristems).

Types of Meristems

- ☆ **Apical Meristem**: Located at the tips of roots and shoots, responsible for primary growth (increase in length).
- ☆ **Lateral Meristem**: Includes the vascular cambium and cork cambium, responsible for secondary growth (increase in girth or thickness).
- ☆ **Intercalary Meristem**: Found in grasses and other monocots, located at the base of leaves or internodes, allowing for regrowth of tissues if the plant is damaged.

Permanent Tissue (Differentiated Tissue)

Permanent tissues consist of cells that have differentiated to perform specific functions. These cells no longer divide but contribute to the plant's structure and function. Permanent tissues are divided into three major categories:

Simple Permanent Tissues

- **Parenchyma**: The most common type of plant tissue, consisting of thin-walled cells that are involved in metabolic functions such as photosynthesis, storage, and secretion. Chlorenchyma (a type of parenchyma) contains chloroplasts and is specialized for photosynthesis.
- **Collenchyma**: Provides flexible support to growing parts of the plant, such as young stems and petioles. Cells have unevenly thickened walls.
- **Sclerenchyma**: Provides rigid support and strength to the plant, often in mature tissues. It includes fibers and sclereids, both of which have thick, lignified cell walls.

Complex Permanent Tissues

- **Xylem**: A vascular tissue responsible for the transport of water and minerals from the roots to the rest of the plant. Xylem consists of tracheids, vessels, fibers, and parenchyma cells.
- **Phloem**: A vascular tissue responsible for the transport of sugars (primarily sucrose) and other organic nutrients from the leaves to other parts of the plant. Phloem contains sieve tubes, companion cells, phloem fibers, and phloem parenchyma.

Epidermal Tissue

The epidermis is the outermost layer of cells covering plant organs. It serves as a protective barrier against water loss, pathogens, and physical damage. The epidermis may include specialized structures such as guard cells (which regulate the opening and closing of stomata), trichomes (hair-like structures that may have a protective or glandular function), and root hairs (for absorption of water and nutrients).

Ground Tissue

- Ground tissue makes up most of the plant's body and includes parenchyma, collenchyma, and sclerenchyma. It is responsible for photosynthesis, storage, and structural support.
- Cortex and pith are examples of ground tissue found in stems and roots. The cortex stores starch and other nutrients, while the pith often stores water or serves as a support structure in the center of the plant.

The Organ Level: Tissues Forming Organs

In plants, tissues combine to form organs, which are specialized structures responsible for carrying out specific physiological functions. The three main organs of a plant are **roots, stems**, and **leaves**, each of which consists of multiple tissues working together.

The Root

- **Function**: The root anchors the plant to the soil, absorbs water and nutrients, and stores food.
- **Tissues**:
    - **Epidermis**: The outermost layer of cells that may form root hairs for increased surface area for absorption.
    - **Cortex**: Ground tissue that stores starch and water.
    - **Vascular Tissue (Xylem and Phloem)**: Found in the central stele (vascular cylinder), responsible for water, mineral, and nutrient transport.
    - **Endodermis**: A specialized layer of cells surrounding the vascular tissue, regulating the flow of water and dissolved minerals into the vascular system.

The Stem

- **Function**: The stem supports the plant's leaves and reproductive organs, transports water, nutrients, and sugars, and may store food.
- **Tissues**:
    - **Epidermis**: The outer protective layer.
    - **Cortex**: Often stores food in the form of starch.
    - **Vascular Tissue**: Composed of xylem and phloem, it is responsible for the upward transport of water and minerals and the downward transport of sugars.
    - **Cambium**: A layer of meristematic cells that produces new xylem and phloem (in woody plants, leading to secondary growth).

The Leaf

- **Function**: The primary organ for photosynthesis, gas exchange, and transpiration.
- **Tissues**:
    - **Epidermis**: Includes the cuticle (a waxy layer) to prevent water loss and **stomata** for gas exchange.
    - **Mesophyll**: The inner tissue of the leaf, consisting of two layers– **palisade mesophyll** (primary site for photosynthesis) and spongy mesophyll (facilitates gas exchange).
    - **Vascular Tissue**: Forms the veins of the leaf, containing both xylem (for water transport) and phloem (for sugar transport).

Plant-Environment Interactions

Plants, as sessile organisms, are deeply intertwined with their environment. Unlike animals, they cannot move to new locations or escape adverse conditions, making their survival and growth highly dependent on their interactions with the surrounding environment. These interactions are multifaceted, involving the exchange of gases, water, and nutrients, as well as responses to various biotic and abiotic factors. Understanding plant-environment interactions is crucial not only for basic plant biology but also for addressing critical global challenges, such as climate change, food security, and environmental conservation.

Abiotic Factors: Environmental Conditions Shaping Plant Life

Plants are influenced by a wide range of abiotic (non-living) factors, which include temperature, light, water, humidity, soil composition, and atmospheric gases. These factors not only determine the physical characteristics and growth patterns of plants but also regulate essential processes like photosynthesis, respiration, and transpiration.

- **Light**: Light is the primary source of energy for plants, driving the process of photosynthesis. The quality, quantity, and duration of light exposure have significant impacts on plant growth. For instance, photoperiod (the length of day and night) regulates flowering in many plants. Shade-tolerant plants have evolved to thrive in low light conditions, while sun-loving plants require direct sunlight for optimal growth. Additionally, light quality, in terms of color spectrum (red, blue light), influences plant behavior through photoreceptors like phytochromes and cryptochromes.
- **Temperature**: Temperature profoundly affects plant metabolism. Each plant species has an optimal temperature range for growth. Cold temperatures can limit enzyme activity and reduce the rate of photosynthesis, while high temperatures can lead to dehydration and heat stress. For instance, high-temperature stress often triggers the production of heat shock proteins that help plants cope with thermal damage. Similarly, freezing temperatures can cause ice formation within plant cells, leading to cell rupture and death. Acclimation to temperature stress is common in plants, with mechanisms like cold hardening or heat shock response pathways.
- **Water Availability**: Water is a critical resource for plants, not only for photosynthesis but also for maintaining turgor pressure, facilitating nutrient transport, and supporting cellular metabolism. Plants face constant challenges related to water availability. In drought conditions, many plants have evolved adaptations such as deep root systems, waxy cuticles to reduce water loss, or the ability to store water in specialized tissues like vacuoles. Conversely, in flooded conditions, plants may adapt by developing aerenchyma tissue (air-filled spaces in roots and stems) to help with gas exchange in anaerobic conditions. Water stress can lead to abscisic acid (ABA) production, a hormone that helps plants close their stomata and conserve water.

- **Soil Quality and Composition**: The soil provides essential nutrients, water, and physical support for plants. Soil texture, structure, pH, and nutrient content greatly affect plant growth. Clay soils, for example, retain water well but may suffer from poor aeration, while sandy soils drain quickly but may lack nutrient-holding capacity. Soil pH affects the availability of nutrients, with many plants preferring slightly acidic to neutral pH (5.5–7). In nutrient-poor soils, plants may develop specialized structures like mycorrhizal associations to enhance nutrient uptake. Soil contamination, such as heavy metal pollution, can also disrupt plant health, causing toxicity and inhibiting growth.
- **Atmospheric Gas Exchange**: Plants exchange gases with the environment through stomata–small pores on the surface of leaves that facilitate the uptake of carbon dioxide (CO_2) for photosynthesis and the release of oxygen (O_2). The concentration of CO_2 in the atmosphere is a critical determinant of photosynthetic efficiency. Rising CO_2 levels due to climate change can enhance photosynthesis in some plants (a phenomenon known as CO_2 fertilization), but this effect may be countered by other factors such as water scarcity or temperature stress. Similarly, the increasing concentration of ozone in the atmosphere can be harmful to plants, causing oxidative stress and reducing photosynthetic efficiency.

Biotic Factors: Plant-Plant and Plant-Animal Interactions

Beyond the abiotic factors, plants are also influenced by various biotic (living) factors, which include interactions with other plants, animals, fungi, and microorganisms. These interactions are dynamic and can have both positive and negative impacts on plant growth, reproduction, and survival.

- **Plant Competition**: In a given environment, multiple plants may compete for the same resources, such as light, water, and nutrients. Plants that grow in dense communities often engage in resource competition, where the most efficient competitors dominate. Allelopathy is another form of competition, where plants release chemicals into the soil to inhibit the growth of neighboring plants. For example, black walnut trees (*Juglans nigra*) produce juglone, a toxic substance that suppresses the growth of nearby plants. On the other hand, some plants form mutualistic relationships with other species, such as the nitrogen-fixing bacteria found in legume roots, which help improve soil nutrient availability.
- **Herbivory**: Herbivores–ranging from insects to larger mammals–can have a significant impact on plant growth and survival. In response to herbivory, plants have developed numerous defensive mechanisms, such as chemical defenses (*e.g.*, toxins, bitter compounds) or physical defenses (*e.g.*, thorns, spines, tough leaves). Many plants also release volatile organic compounds (VOCs) to signal nearby plants to activate

defense mechanisms, such as producing defensive proteins or attracting herbivore predators (a phenomenon known as indirect defense).

- **Pollination**: Successful reproduction in many plants depends on pollination–the transfer of pollen from male to female reproductive organs. Plants rely on a variety of pollinators, including insects, birds, bats, and even wind. Flower morphology and coloration often evolve to attract specific pollinators. For example, bees are attracted to blue or purple flowers, while hummingbirds prefer red flowers. Pollinators are not only critical for plant reproduction but also for maintaining ecosystem biodiversity.
- **Mycorrhizal and Symbiotic Relationships**: Plants often engage in beneficial symbiotic relationships with fungi and bacteria. One of the most well-known examples is the association between plant roots and mycorrhizal fungi, where the fungus helps the plant absorb nutrients, especially phosphorus, in exchange for carbohydrates produced by the plant through photosynthesis. Similarly, plants like legumes engage in symbiosis with rhizobium bacteria, which fix atmospheric nitrogen into a form that plants can use for growth. These symbiotic relationships play a significant role in enhancing plant nutrient acquisition and improving ecosystem functioning.

Climate Change and Global Environmental Changes

The relationship between plants and their environment is increasingly shaped by global environmental changes, particularly climate change. Rising temperatures, altered precipitation patterns, increasing CO_2 levels, and more frequent extreme weather events are already having profound effects on plant growth and distribution. For instance, warmer temperatures can shift the growing seasons, while droughts can limit water availability and reduce agricultural productivity. On the other hand, elevated CO_2 may enhance photosynthesis in some plants, but this benefit may be mitigated by other limiting factors such as nutrient deficiencies or heat stress. Moreover, shifts in plant distribution due to changing climate patterns can alter entire ecosystems, as plants interact with animals, insects, and other plants in complex ways. Understanding how plants respond to environmental changes is essential for developing strategies to mitigate the impacts of climate change on agriculture and natural ecosystems.

Plant-environment interactions are the result of complex and dynamic relationships that allow plants to survive, grow, and reproduce in a wide variety of environments. Both abiotic factors like light, temperature, water, and soil composition, and biotic factors like competition, herbivory, and symbiosis, play a critical role in shaping plant biology. These interactions are not static; they change over time in response to environmental fluctuations and evolutionary pressures.

As human activities continue to alter the global environment, understanding plant-environment interactions becomes increasingly important for improving agricultural productivity, conserving biodiversity, and addressing the challenges posed by climate change. Through research and careful management, we can better understand and protect the plants that form the foundation of ecosystems and human livelihoods.

Chapter 2

Water and Mineral Nutrition

Introduction

Water is an important factor for plant growth as it helps to fulfill all the vital activities of plants. Water is essential for photosynthesis, respiration, absorption of minerals and nutrients, metabolism and even to maintain the soil temperature too. Beside this, water is also important in various other processes too, as it helps in the germination of seeds and in the process of transpiration *etc.* Water helps a plant by transporting nutrients through the roots. Nutrients are drawn from the soil and used by the plant. Without enough water in the cells, the plants droop so water helps a plant stand. Water carries the dissolved sugar and other nutrients through the roots. Plants absorb water through their entire surface- roots, stems and leaves. However, the majority of water is absorbed by root hairs. To maintain the level of water inside the plant cells, it is necessary, to loss excess water from plant cells either in the form of evaporation or through transpiration. Evaporation of water from leaves is primarily controlled by stomata, sometimes lenticels and pores also helps in this process. This shows that, plants have a strong and significant relationship with water. Plant water relation means plants control the hydration of their cells including the collection of water from the soil, its transport within the plants and its loss by evaporation from the leaves. Transpiration also includes a process called guttation, which is the loss of water in liquid form from the uninjured leaf or stem of the plant principally through water stomata known as hydathodes. Studies have revealed that about 10 percent of the moisture found in the atmosphere is releases by plants through transpiration.

Importance of Water to Plant Life

Water is most important and prime factor for life processes, as life itself has been originated in an aqueous environment. In course of evolution, it became fully dependent on water in a number of ways. Thus, we can say that water is the liquid of life or elixir of life. In general water is essential for life and is the main constituent of the protoplasm comprising 90 to 95 per cent of its total weight. In absence of water, protoplasm becomes inactive and even killed. Water is a source of hydrogen atoms for the reduction of carbon dioxide in the reaction of photosynthesis and as mentioned earlier that water helps to fulfill different vital activities. Beside this water present in the vacuoles which helps in maintaining the turgidity of cells which is essential for proper activities of life. Due to absorption of water the cell becomes turgid. The turgidity of cell helps in the elongation of cells resulting in growth. It is a well- known fact that the availability of water during winter and summer season is different and is the reason of formation of annual rings in higher plants. In summer the turgidity is less and hence formation of small cells, comparatively to winter when due to higher turgidity, formation of larger cells takes place. In nature we have different type plants *e.g.* aquatic, terrestrial, halophytes, xerophytes *etc.* and different plants absorb water in different ways. While orchids, absorb moisture directly from the atmosphere and not from soil. Land plants get their water supply from soil which serves as the source of water and minerals to them. The way in which water from soil enters into roots, particularly to root xylem is called Mechanism of water absorption.

Figure 2.1. Different States of Matter.

Water Movement in Plants

Water movement in plants is a fundamental and vital process that underpins the life and function of plants. It is central to nearly all physiological activities, including nutrient uptake, photosynthesis, temperature regulation, and maintaining cellular structure. Plants, as immobile organisms, rely on highly specialized mechanisms to move water from the soil, through their roots, upward through the stem, and ultimately to the leaves and other parts of the plant. This movement of water is

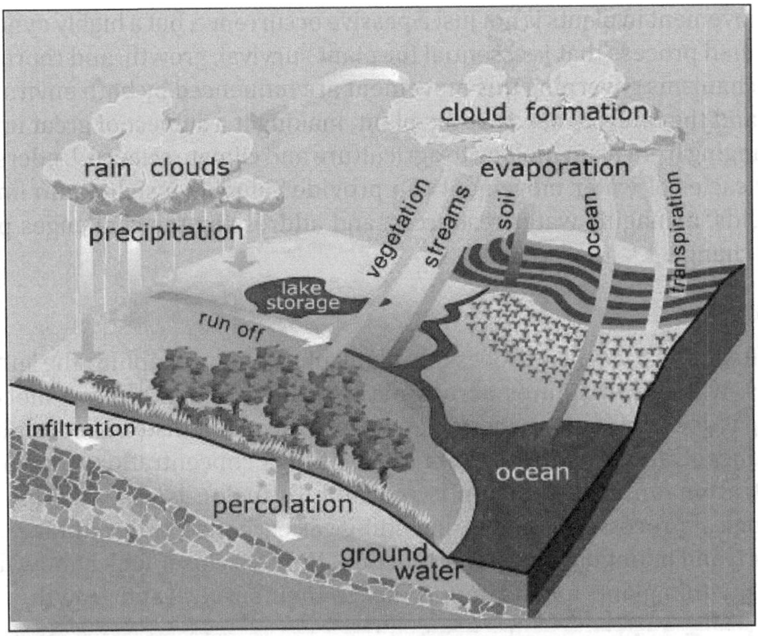

Figure 2.2. Water Cycle in the Atmosphere.

essential not only for sustaining plant cells but also for facilitating the transport of minerals and nutrients that are critical for growth and development. Water movement in plants occurs through a combination of passive and active processes, and it involves various complex structures within the plant. It begins with the absorption of water by the roots, primarily through osmosis, as water moves from areas of high water potential in the soil to areas of lower water potential within the root cells. Once absorbed, water travels through the vascular tissues, particularly the xylem, a specialized system of vessels that form a continuous network running from the roots to the leaves. This upward movement of water is driven by a variety of forces, including transpiration (the evaporation of water from the leaf surfaces), cohesion (the attraction between water molecules), and adhesion (the attraction between water molecules and the walls of the plant's vascular system). Transpiration plays a particularly critical role in driving water through plants, as the loss of water vapor from the stomata of the leaves creates a negative pressure within the plant's vascular system, essentially "pulling" water upward from the roots. This upward pull is assisted by the cohesive and adhesive properties of water, which help it move against gravity. At the same time, plants regulate this process through mechanisms such as the opening and closing of stomata, balancing the need for water conservation with the necessity of gas exchange for photosynthesis. The ability of plants to efficiently manage water movement is a key factor in their ability to thrive in diverse environmental conditions, and disruptions to this process–such as drought or disease–can significantly impact plant health and productivity. Thus, water movement in plants is not just a passive occurrence, but a highly dynamic and finely tuned process that is essential for plant survival, growth, and reproduction. The mechanisms governing this movement are influenced by both environmental factors and the internal needs of the plant, making it a subject of great interest in fields ranging from plant biology to agriculture and climate science. Understanding the intricacies of water movement can provide valuable insights into improving crop yields, managing water resources, and addressing the challenges posed by climate change.

Diffusion

Diffusion is a critical process in plant physiology, enabling the movement of gases, water, and solutes across plant tissues. It is a passive mechanism, meaning it does not require energy input from the plant. Instead, molecules move along concentration gradients–from areas of high concentration to areas of low concentration–until equilibrium is reached. Diffusion is essential for various physiological processes in plants, including gas exchange, water uptake, nutrient transport, and maintaining cellular function. Here's a closer look at how diffusion operates within plants and its importance in their survival and growth.

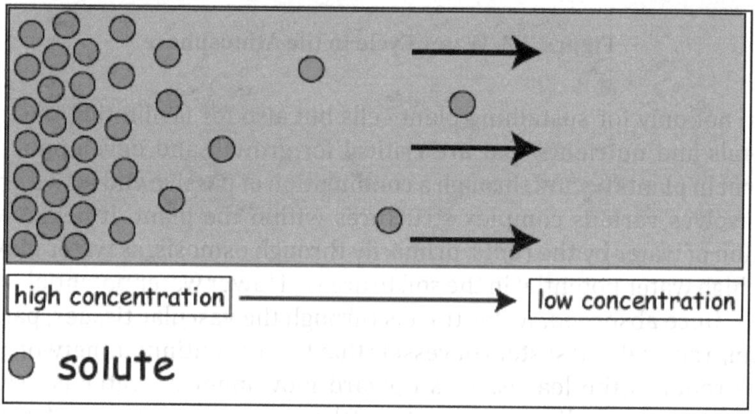

Figure 2.3. Solute Transport is from the Left to the Right, Movement of the Solutes is Due to the Concentration Gradient (dC/dx).

Processes Involving Diffusion in Plants

Gas Exchange (Respiration and Photosynthesis)

Gas exchange in plants relies heavily on diffusion, particularly through stomata, small pores on the surface of leaves and stems. These pores allow gases like carbon dioxide (CO_2) and oxygen (O_2) to diffuse in and out of the plant tissues.

- **During Photosynthesis:** In the presence of light, plants take in carbon dioxide (CO_2) from the air through the stomata. CO_2 diffuses into the leaf, where it is used by chloroplasts in photosynthetic cells to produce glucose and oxygen. Oxygen, a byproduct of photosynthesis, diffuses out of the plant and into the surrounding atmosphere.

☆ **During Respiration:** In the dark or when the plant is respiring, oxygen from the air diffuses into the plant for cellular respiration, where it is used to break down glucose for energy. Carbon dioxide, produced as a waste product, diffuses out of the plant through the stomata.

The process of diffusion ensures that plants can exchange gases efficiently with the surrounding environment, which is essential for maintaining metabolic activities such as photosynthesis and respiration.

Water Movement (Osmosis)

Water movement in plants is also driven by diffusion, though the specific process involved is osmosis–a type of diffusion involving water molecules. Water moves through plant cells and tissues from areas of high-water potential to areas of lower water potential.

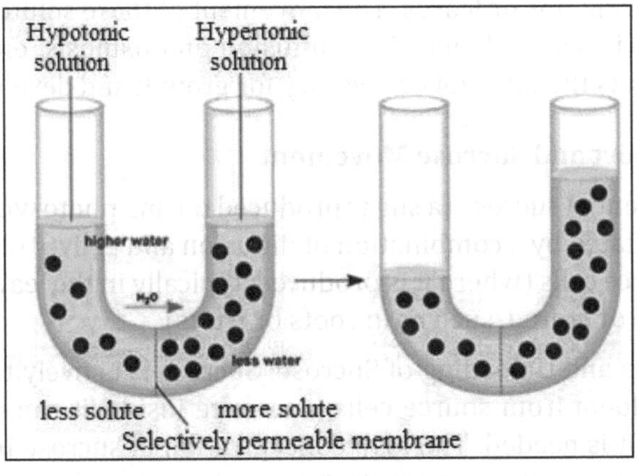

Figure 2.4. Demonstration of Osmosis Phenomenon.

☆ **Water Uptake in Roots:** Water enters plant roots from the surrounding soil primarily through osmosis. The concentration of water is higher in the soil than in the root cells, so water diffuses into the root through the cell membranes. The movement of water helps maintain the plant's turgor pressure, which is essential for keeping cells firm and ensuring that nutrients can be transported throughout the plant.

☆ **Transpiration and Water Loss:** Water in the plant's vascular system (specifically in the xylem) moves upwards due to a combination of capillary action, cohesion, and transpiration pull. Transpiration is the evaporation of water from the plant's leaves through the stomata. As water vapor diffuses out of the stomata, it creates a negative pressure that pulls more water upward from the roots, maintaining the flow of water through the plant. This continuous water movement is critical for transporting nutrients and minerals from the soil into the plant.

Nutrient and Mineral Uptake

Diffusion also plays a vital role in the uptake of minerals and nutrients from the soil into the plant. The concentration of minerals in the soil is typically higher than inside the root cells, so they move into the plant tissues via diffusion. For instance:

- Minerals in Soil: Essential nutrients like potassium (K^+), nitrogen (in the form of nitrate ions, NO_3^-), and phosphates are often absorbed by roots through diffusion. The plant's root system maintains a higher concentration of solutes (minerals) within its cells compared to the surrounding soil, which drives the diffusion of minerals into the plant roots.
- Movement of Nutrients in the Plant: Once inside the roots, minerals are transported through the plant's vascular system to areas of need, such as growing tips or leaves. The movement of these solutes through plant tissues is also influenced by diffusion and osmosis, ensuring that the plant gets the nutrients necessary for growth and development.

Phloem Transport and Sucrose Movement

The movement of sucrose (a sugar produced during photosynthesis) through the plant is facilitated by a combination of diffusion and active transport. Sucrose moves from source cells (where it is produced, typically in the leaves) to sink cells (where it is used or stored, such as in roots or fruits).

- Loading and Unloading of Sucrose: Sucrose is actively transported into the phloem from source cells, and once inside, it can diffuse to areas where it is needed. The high concentration of sucrose in the phloem at the source creates a concentration gradient that allows sucrose to move passively towards sinks, where the concentration is lower.

Although active transport is involved in loading sucrose into the phloem, the overall movement of nutrients and sugars in the plant is heavily influenced by diffusion, especially as the sugars move across membranes and into surrounding tissues.

Factors Affecting Diffusion in Plants

The rate of diffusion in plants is influenced by several factors:

- **Concentration Gradient:** A steeper concentration gradient (a larger difference between the concentrations of a substance on either side of a membrane) leads to a faster rate of diffusion. For example, if the concentration of carbon dioxide inside a leaf is lower than the concentration in the atmosphere, CO_2 will diffuse into the leaf more rapidly.

☆ **Temperature:** Higher temperatures increase the kinetic energy of molecules, which speeds up the movement of particles. In plants, warmer temperatures can enhance the rate of diffusion for gases like CO_2 and O.

☆ **Surface Area:** A larger surface area allows for more molecules to diffuse across. For example, the vast network of stomata in plant leaves increases the surface area available for gas exchange, facilitating the diffusion of gases into and out of the plant.

☆ **Membrane Permeability:** The permeability of membranes to different substances also influences diffusion. For example, the lipid bilayer of plant cell membranes is selectively permeable, allowing gases like CO_2 to diffuse through more easily, while larger or polar molecules may require assistance from transport proteins.

☆ **Water Potential:** Water diffusion in plants (osmosis) depends on water potential, which is influenced by solute concentration and physical pressure. Water tends to move from areas of higher water potential to lower water potential, influencing processes like water uptake in roots and the movement of water up through the plant.

Diffusion is a critical mechanism in plant physiology, responsible for essential processes such as gas exchange, water uptake, nutrient absorption, and internal transport. By allowing molecules to move along concentration gradients, diffusion enables plants to take in vital substances like carbon dioxide and water while expelling waste products like oxygen and carbon dioxide. Additionally, diffusion helps maintain turgor pressure, supports nutrient transport through the phloem, and ensures efficient gas exchange during photosynthesis and respiration. The efficiency of diffusion in plants is influenced by various factors, including concentration gradients, temperature, surface area, and membrane properties. In this way, diffusion is integral to maintaining the health, growth, and function of plants.

Bulk Flow

Bulk flow in plant physiology refers to the mass movement of water, nutrients, and other substances through plant tissues, driven by pressure gradients rather than diffusion. This process occurs primarily in the plant's vascular system–specifically in the xylem and phloem–and is essential for transporting water, minerals, sugars, and hormones over long distances within the plant. In the xylem, bulk flow is driven by the transpiration pull–the evaporation of water from the stomata of leaves creates a negative pressure that pulls water upward from the roots. This process is assisted by cohesion (water molecules sticking together) and adhesion (water molecules sticking to the walls of xylem vessels), enabling the movement of water and dissolved minerals from the roots to the leaves. In the phloem, bulk flow occurs through a process called pressure flow or mass flow, where sugars (primarily sucrose) are actively loaded into the phloem at the source (such as the leaves), creating a high osmotic pressure. This pressure drives the flow of the phloem sap toward sinks (such as roots or fruits), where the sugars are used or stored. Bulk flow is crucial for maintaining the plant's hydration, nutrient transport, and overall growth, as it ensures that essential substances are moved efficiently across the plant's vast internal network of tissues.

Water Uptake by Roots

Water uptake by plant roots is a vital process for maintaining hydration, nutrient transport, and overall plant health. It occurs primarily through the root hairs, which are thin, specialized extensions of the epidermal cells of the roots. The process of water uptake is largely driven by osmosis, a passive movement of water molecules from an area of higher water potential (the soil) to an area of lower water potential (inside the root cells). This process ensures that water, along with essential dissolved minerals, enters the plant system to support various physiological functions.

Mechanisms of Water Uptake

Osmosis

The main mechanism of water uptake in roots is osmosis, which occurs when water moves from the soil, where water potential is high, into the root cells, where the water potential is lower due to the presence of dissolved solutes (such as minerals and sugars). Water molecules diffuse across the selectively permeable cell membranes of root hair cells and into the interior of the root. The movement continues as long as there is a difference in water potential between the soil and the root cells.

Apoplast and Symplast Pathways

Once water enters the root hair cells, it can move toward the vascular tissue (xylem) through two main pathways:

- **Apoplast Pathway**: Water moves through the cell walls and the spaces between cells without crossing any cell membranes. This is a faster route but is blocked at the Casparian strip–a band of suberin (waxy material) in the endodermis (the innermost layer of the root cortex) that forces water to move symplastically.
- **Symplast Pathway**: Water moves through the cytoplasm of root cells via the plasmodesmata, small pores that connect adjacent cells. This pathway is slower than the apoplast pathway but allows for more selective movement of water and solutes.

Root Pressure

Root pressure is a phenomenon where water is pushed into the xylem vessels of the roots due to the accumulation of water in the root cortex. This occurs when there is a high concentration of solutes in the root cells, which draws water in by osmosis. The pressure generated can help push water up to the stem, although it is generally not enough to move water to the top of tall trees. Root pressure is more noticeable in certain plants during the night, when transpiration (the loss of water through leaves) is low.

Active Transport of Minerals

While water enters the roots passively through osmosis, the uptake of minerals such as potassium (K^+), nitrate (NO_3^-), and phosphate (PO_4^{3-}) is generally an active process. This requires energy, as ions are transported against their concentration gradients by membrane-bound protein pumps in the root cell membranes. The movement of minerals into the root cells lowers their water potential, further driving the influx of water from the surrounding soil.

Factors Affecting Water Uptake by Roots

Several factors influence the rate and efficiency of water uptake by roots:

- ☆ **Soil Water Availability**: The amount of water in the soil directly affects how much water can be absorbed. Drought conditions or compacted soil can reduce the availability of water, limiting root uptake.
- ☆ **Root Surface Area**: A larger root surface area, particularly through the presence of root hairs, increases the contact between the root and the soil, allowing for more efficient water absorption.
- ☆ **Soil Composition**: The structure and texture of the soil impact water retention and movement. Soils with good porosity and good drainage allow for optimal water uptake, while heavy clay soils or compacted soils can hinder water movement and root growth.
- ☆ **Water Potential Gradient**: The greater the difference in water potential between the soil and the root, the more water will flow into the plant. Factors such as soil salinity, mineral content, and plant metabolic activity influence this gradient.
- ☆ **Temperature**: Warmer temperatures generally increase the rate of water uptake by increasing the kinetic energy of water molecules and enhancing the functioning of enzymes involved in metabolic processes that drive nutrient and water absorption.

Role of Water in Plant Growth and Development

The water taken up by the roots serves several key functions:

- ☆ **Transport of Nutrients**: Water acts as the medium in which dissolved minerals and nutrients are transported throughout the plant, especially through the vascular system (xylem and phloem).

- **Turgor Pressure**: Water is crucial for maintaining turgor pressure, which helps keep plant cells rigid and supports plant structure. Turgor pressure also drives cell expansion during growth.
- **Photosynthesis**: Water is a key component in photosynthesis, where it is split to release oxygen, electrons, and protons that are essential for the production of glucose and energy for the plant.
- **Temperature Regulation**: Water helps regulate the plant's internal temperature through evaporation and transpiration, preventing overheating and promoting the movement of water from the roots to the leaves.

Water uptake by roots is a fundamental process for plant health and function, facilitated by osmosis and influenced by various factors such as soil conditions, root structure, and the plant's metabolic activities. Through osmosis, water enters the root cells and moves into the plant's vascular system, where it is transported to various tissues. This movement is crucial for maintaining hydration, supporting nutrient transport, and enabling key processes like photosynthesis and growth. Understanding how water is absorbed by roots helps to explain how plants adapt to varying environmental conditions and optimize their water use for survival and productivity.

Transpiration: Mechanism and Functions

Transpiration is the process by which water is lost from the plant to the atmosphere, primarily through evaporation from the stomata (small pores) on the leaves and stems. While transpiration is often associated with water loss, it is also a crucial physiological process that facilitates a range of functions necessary for plant survival and growth. It plays an integral role in the plant's water and nutrient management system, influencing processes like water uptake, cooling, and gas exchange. Transpiration is not a passive process but one that is regulated and affected by environmental conditions, the plant's metabolic activity, and the structure of its leaves.

Mechanism of Transpiration

The process of transpiration is driven by the water potential gradient between the inside of the leaf and the surrounding environment. Water moves from areas of high water potential (inside the plant, in the root system and vascular tissues) to areas of low water potential (in the atmosphere) through various mechanisms.

1. **Water Movement Through the Plant**

   - Water is absorbed by the roots from the soil through osmosis and moves upward through the plant via the xylem. This upward movement of water is facilitated by a combination of forces, including cohesion (the attraction between water molecules), adhesion (the attraction of water molecules to the walls of the xylem vessels), and capillary action.

- ☆ As water travels through the xylem, it moves toward the leaves, where it can exit through the stomata. In the leaves, water is present in the intercellular spaces, and it evaporates into the atmosphere via the stomatal pores.

2. Evaporation and Stomatal Regulation

- ☆ Transpiration begins when water evaporates from the mesophyll cells (the inner tissues of the leaf). This water vapor then accumulates in the air spaces within the leaf and diffuses out through the stomata, which are controlled by guard cells.
- ☆ The stomata are tiny openings found on the surface of leaves and stems, and their opening and closing are regulated by guard cells. When the guard cells take up water and become turgid, they open the stomata. When the guard cells lose water, they become flaccid, causing the stomata to close.
- ☆ The rate of transpiration is influenced by the number of stomata, their size, and how open they are. Plants can regulate their stomatal opening to minimize water loss during drought conditions or maximize gas exchange during active photosynthesis.

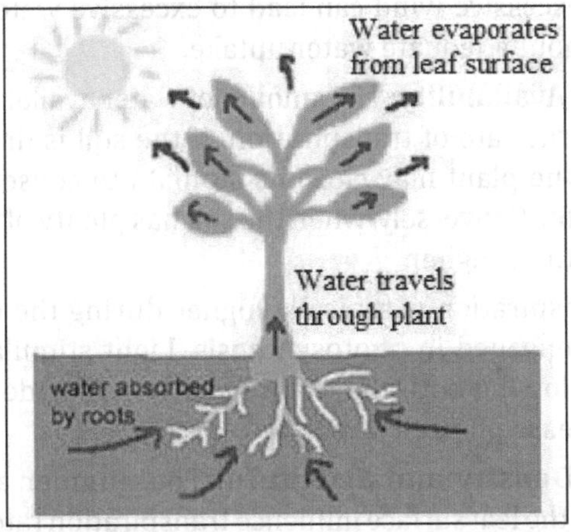

Figure 2.5. Stomatal Transpiration.

3. Role of Transpiration Pull

- ☆ As water evaporates from the leaf, it creates a negative pressure or tension in the leaf's water column, which is transmitted through the xylem all the way to the roots. This is known as the transpiration pull, and it helps draw water from the soil into the roots and up through the plant.

☆ The cohesion of water molecules (due to hydrogen bonding) ensures that the water column remains intact as it moves through the plant, facilitating a continuous flow from the roots to the leaves.

Factors Affecting Transpiration

Several environmental and internal factors regulate the rate of transpiration in plants:

1. **Humidity**: Transpiration occurs more rapidly when the surrounding air is dry (low humidity) because the water vapor in the leaf's intercellular spaces will diffuse more readily into the atmosphere. High humidity slows transpiration as the concentration gradient between the inside of the leaf and the external environment decreases.
2. **Temperature**: Higher temperatures increase the rate of evaporation, leading to an increased rate of transpiration. Warm temperatures also increase the plant's metabolic activity, including photosynthesis, which requires more water for both nutrient transport and cooling.
3. **Wind**: Wind increases transpiration by removing the saturated layer of air around the stomata, thereby increasing the concentration gradient and promoting the diffusion of water vapor from the leaf to the atmosphere. However, excessive wind can lead to excessive water loss, especially in plants without adequate water uptake.
4. **Soil Water Availability**: The amount of water available in the soil directly influences the rate of transpiration. If the soil is dry and water uptake is limited, the plant may close its stomata to conserve water, reducing transpiration. Conversely, when the soil has plenty of water, transpiration rates tend to be higher.
5. **Light**: Transpiration is typically higher during the day when the plant is actively engaged in photosynthesis. Light stimulates the opening of stomata, allowing both the entry of carbon dioxide for photosynthesis and the release of water vapor.
6. **Stomatal Density and Structure**: The number and distribution of stomata on the leaf surface influence transpiration rates. Plants adapted to dry environments (xerophytes) often have fewer stomata or stomata that are sunken to reduce water loss. Conversely, plants in moist environments (mesophytes) typically have more stomata to facilitate gas exchange.

Functions of Transpiration

Although transpiration results in the loss of water, it serves several vital functions in plants:

☆ **Water and Nutrient Transport**: Transpiration creates a pulling force (the transpiration pull) that drives the upward movement of water through the plant's vascular system, facilitating the transport of essential nutrients and minerals from the soil to the leaves and other growing tissues.

- **Cooling**: Transpiration helps cool the plant by dissipating heat. As water evaporates from the leaf surface, it absorbs heat from the plant, reducing the temperature of leaf tissues. This process is similar to sweating in animals and helps prevent overheating, particularly in hot or sunny conditions.
- **Turgor Pressure Maintenance**: The continuous uptake of water due to transpiration helps maintain turgor pressure in plant cells, which is crucial for maintaining cell rigidity and structure. Turgor pressure helps the plant stand upright and allows for growth, as it supports cell expansion.
- **Gas Exchange**: Transpiration facilitates gas exchange in plants. The opening of stomata, which is necessary for transpiration, also allows for the intake of carbon dioxide (CO_2) for photosynthesis and the release of oxygen (O_2), a byproduct of photosynthesis. The rate of transpiration is closely linked to photosynthetic activity.
- **Mineral and Water Regulation**: Transpiration contributes to the regulation of mineral and water balance in the plant. The continuous flow of water through the plant ensures that excess salts and waste products are removed, and that the plant can maintain optimal hydration for growth.

Types of Transpiration

- **Cuticular Transpiration**: This occurs through the cuticle, the waxy layer covering the surface of leaves and stems. Though it accounts for a smaller fraction of total transpiration, it plays a role in plants with thicker cuticles, particularly in arid environments.

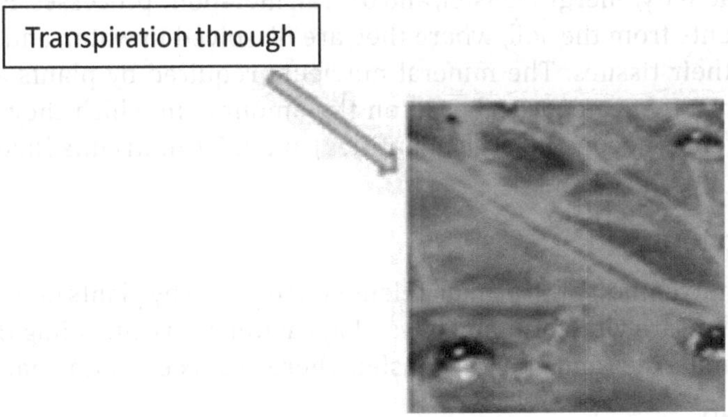

☆ **Stomatal Transpiration**: The majority of water loss occurs through the stomata. This is the primary pathway for transpiration in most plants and is highly regulated based on environmental factors and the plant's water status.

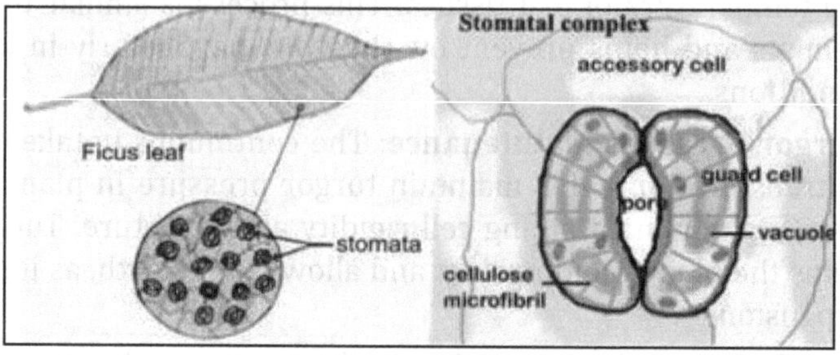

☆ **Lenticular Transpiration**: This type of transpiration occurs through small openings in the plant's bark, called **lenticels**. It is more common in woody plants and accounts for a small fraction of overall water loss.

Mineral Elements in Plant Nutrition

Mineral elements are essential for the growth and development of plants. They serve a variety of critical functions, including promoting structural integrity, enzymatic activity, energy transfer, and overall metabolic processes. Plants absorb these elements from the soil, where they are dissolved in water, and incorporate them into their tissues. The mineral nutrients required by plants are typically classified into two categories based on the amounts in which they are needed: macronutrients (needed in large quantities) and micronutrients (needed in trace amounts).

Macronutrients

Macronutrients are the primary elements required by plants in relatively large amounts. They are vital for many physiological functions, including the synthesis of organic molecules and energy transfer. There are six essential macronutrients:

Nitrogen (N)

☆ **Function**: Nitrogen is a key component of amino acids, proteins, and chlorophyll. It is essential for growth and development, influencing both vegetative growth and photosynthesis. Nitrogen is also involved in the production of nucleic acids (DNA and RNA) and enzymes.

- ☆ **Deficiency Symptoms**: A nitrogen deficiency typically results in yellowing of older leaves (chlorosis), poor growth, and stunted development because nitrogen is a mobile nutrient and will be redistributed from older to newer tissues.

Phosphorus (P)

- ☆ **Function**: Phosphorus is a component of ATP (adenosine triphosphate), which is essential for energy transfer in cells. It is also a key element in the formation of DNA, RNA, and phospholipids, which are crucial for cell membranes. Phosphorus supports root development and flowering.
- ☆ **Deficiency Symptoms**: Phosphorus deficiency leads to dark green leaves with purple or red coloration, particularly in younger leaves. The plant may also show poor root development and delayed flowering.

Potassium (K)

- ☆ **Function**: Potassium helps in maintaining osmotic balance and regulating the opening and closing of stomata, which controls water loss and gas exchange. It is also involved in enzyme activation, protein synthesis, and carbohydrate metabolism.
- ☆ **Deficiency Symptoms**: Potassium deficiency results in yellowing of leaf edges (marginal chlorosis), especially in older leaves, and poor root and fruit development. The plant may also exhibit necrotic spots and a weakened ability to resist disease.

Calcium (Ca)

- ☆ **Function**: Calcium is crucial for maintaining cell wall structure, cell division, and signal transduction. It also stabilizes the structure of plant membranes and is involved in the regulation of enzymatic activity.
- ☆ **Deficiency Symptoms**: Calcium deficiency often leads to necrosis (death) of young tissue, particularly at the tips of roots and leaves (commonly called "tip burn"). It may also cause blossom-end rot in fruits like tomatoes.

Magnesium (Mg)

- ☆ **Function**: Magnesium is the central atom in the chlorophyll molecule and is crucial for photosynthesis. It also plays a role in activating enzymes involved in the synthesis of nucleic acids and proteins.
- ☆ **Deficiency Symptoms**: Magnesium deficiency causes interveinal chlorosis (yellowing between veins) in older leaves, which can lead to the death of older plant tissues. Leaves may also become bronzed or purple.

Sulfur (S)

- **Function**: Sulfur is a component of certain amino acids (such as cysteine and methionine) and vitamins. It is also involved in the synthesis of coenzyme A, which is necessary for the production of energy-rich compounds.
- **Deficiency Symptoms**: Sulfur deficiency leads to yellowing of young leaves, particularly between the veins (similar to nitrogen deficiency), and reduced growth and vigor.

Micronutrients

Micronutrients, though required in much smaller quantities, are equally essential for plant health. They often act as cofactors for enzymes and are involved in a variety of metabolic processes. The key micronutrients include:

Iron (Fe)

- **Function**: Iron is essential for chlorophyll synthesis and is involved in electron transport during photosynthesis and respiration. It also plays a role in the activation of various enzymes.
- **Deficiency Symptoms**: Iron deficiency results in interveinal chlorosis (yellowing between veins) in young leaves. In severe cases, the leaves may turn completely yellow.

Manganese (Mn)

- **Function**: Manganese is involved in photosynthesis, particularly in the water-splitting reaction that occurs in the chloroplasts. It also plays a role in the synthesis of certain enzymes and in nitrogen assimilation.
- **Deficiency Symptoms**: Manganese deficiency leads to interveinal chlorosis and may result in necrotic spots on older leaves. It also affects photosynthetic efficiency.

Zinc (Zn)

- **Function**: Zinc is essential for the synthesis of enzymes involved in protein synthesis, growth regulation, and cell division. It also plays a role in the production of **auxins**, which regulate growth.
- **Deficiency Symptoms**: Zinc deficiency results in stunted growth, shortened internodes, and chlorosis of young leaves. Leaf tips may turn white or develop mottled patterns.

Copper (Cu)

- **Function**: Copper is involved in photosynthesis and respiration as it helps in electron transport. It is also crucial for the functioning of certain enzymes and for the formation of lignin in plant cell walls.

- ☆ **Deficiency Symptoms**: Copper deficiency causes stunted growth, chlorosis, and necrosis of leaf tips. It can also lead to poor flowering and fruiting.

Boron (B)

- ☆ **Function**: Boron is essential for the formation of cell walls and plays a role in sugar transport, cell division, and the formation of pollen tubes during fertilization.
- ☆ **Deficiency Symptoms**: Boron deficiency leads to discolored and dead tips on young leaves, blossom-end rot, and poor fruit development.

Molybdenum (Mo)

- ☆ **Function**: Molybdenum is involved in the nitrogen metabolism of plants, including the reduction of nitrate to ammonium, a process crucial for protein synthesis.
- ☆ **Deficiency Symptoms**: Molybdenum deficiency causes yellowing of leaves, particularly in older tissues, and affects nitrate assimilation, leading to poor growth.

Chlorine (Cl)

- ☆ **Function**: Chlorine is important for osmotic regulation, ion balance, and is involved in the photosynthetic process, particularly in the splitting of water during light reactions.
- ☆ **Deficiency Symptoms**: Chlorine deficiency is rare but can cause wilting, chlorosis, and leaf curling, particularly in young leaves.

Functions of Mineral Elements in Plants

- ☆ **Structural Components**: Many minerals, such as calcium, magnesium, and potassium, are crucial for maintaining the structure and stability of plant cells. Calcium, for example, is involved in the formation of cell walls, while magnesium is central to the structure of chlorophyll.
- ☆ **Enzyme Activation**: Many minerals act as cofactors for enzymes, facilitating biochemical reactions within the plant. Iron, for instance, is required for electron transport chains during photosynthesis and respiration.
- ☆ **Energy Transfer and Storage**: Phosphorus and sulfur are involved in energy storage and transfer. Phosphorus, through ATP, is essential for energy release during cellular processes, while sulfur is involved in the synthesis of essential coenzymes like coenzyme A.
- ☆ **Water and Nutrient Transport**: Potassium helps regulate the stomatal opening and closing, which controls water loss through transpiration

and maintains ion balance within the plant cells. Magnesium is involved in the transport of sugars.

- ☆ **Photosynthesis and Respiration**: Several minerals, especially magnesium, iron, and manganese, are essential for the chlorophyll structure and the electron transport processes during photosynthesis. Copper, zinc, and molybdenum are also involved in respiratory processes.

Soil-Plant Relationships and the Role of Nutrient Uptake

The relationship between soil and plants is fundamental to plant growth and development, influencing the plant's access to essential nutrients, water, and support. Soils act as both the physical and chemical medium from which plants draw the nutrients necessary for life. This symbiotic relationship is not just about the physical space provided by the soil for the plant's roots to anchor and grow but also about the dynamic interaction between the roots and the soil's mineral and organic components, which support nutrient uptake and water absorption. A healthy soil-plant relationship is crucial for maximizing plant growth, yield, and resilience to environmental stressors.

Soil Composition and its Effect on Plant Growth

Soil is a complex mixture of organic matter, minerals, water, air, and living organisms. Its composition greatly influences the availability of nutrients, the root environment, and the plant's overall access to essential resources. Soils can vary widely in terms of texture, structure, pH, and nutrient content, which all affect the plant's ability to thrive.

- ☆ **Soil Texture**: The texture of the soil, determined by the relative proportions of sand, silt, and clay particles, directly affects its ability to retain and drain water, and its capacity to hold nutrients. Sandy soils tend to drain quickly and may not retain enough water or nutrients for plant growth. In contrast, clay soils hold water and nutrients well but may become compacted, reducing root growth and oxygen availability. Loamy soils, which have a balanced texture, provide the best environment for most plants as they offer adequate drainage and good nutrient retention.
- ☆ **Soil Structure**: The arrangement of soil particles into aggregates (clumps) is known as soil structure. Good soil structure is essential for root penetration, water retention, and the movement of gases like oxygen and carbon dioxide. A well-structured soil has enough pore spaces to allow air and water to circulate freely, which is critical for root health and nutrient uptake.
- ☆ **Soil pH**: The pH level of the soil influences nutrient availability. Most plants thrive in soils with a slightly acidic to neutral pH (pH 6-7). If the pH is too high or too low, certain nutrients may become less available to plants. For example, in acidic soils (low pH), phosphorus, calcium, and

magnesium become less available, while in alkaline soils (high pH), iron, manganese, and zinc may be deficient.

Nutrient Uptake Mechanism

Plants take up essential nutrients from the soil primarily through their roots, which are in constant contact with the soil solution, a mixture of water and dissolved minerals. Nutrient uptake is a complex process influenced by various factors such as soil conditions, plant health, and environmental factors. There are several key mechanisms by which plants absorb nutrients from the soil:

- **Passive Transport (Diffusion):** The primary mechanism by which plants absorb nutrients like water and minerals is through diffusion. Nutrients naturally move from areas of higher concentration in the soil solution to areas of lower concentration within the root cells. For example, potassium and nitrate ions move into the root cells from the surrounding soil by diffusion. This process requires no energy expenditure from the plant.
- **Active Transport:** While some nutrients can diffuse passively into plant cells, others, particularly those that are required in larger quantities or in low concentrations in the soil, are actively transported. This process involves energy (ATP) and the action of specific membrane transport proteins. For example, the uptake of phosphates, sulfates, and minerals like calcium often occurs through active transport mechanisms that move ions against their concentration gradients.
- **Ion Exchange:** The process of ion exchange plays a significant role in nutrient uptake. Plants can release hydrogen ions (H^+) from their roots into the soil in exchange for essential cations like potassium (K^+) and calcium (Ca^{2+}). This helps to maintain nutrient balance in the rhizosphere (root zone) and ensures that nutrients are available for uptake.
- **Mycorrhizal Symbiosis:** Many plants form a mutualistic relationship with fungi called mycorrhizae, which extend the root system and enhance the plant's ability to absorb nutrients, especially phosphorus, which can be limiting in many soils. In exchange for sugars produced by the plant during photosynthesis, the mycorrhizal fungi help to break down organic matter in the soil and assist in nutrient absorption.
- **Root Exudates:** Plants also release a variety of organic compounds (such as sugars, amino acids, and organic acids) into the soil. These root exudates help in several ways: they can solubilize nutrients that are otherwise unavailable, attract beneficial microorganisms (such as nitrogen-fixing bacteria), and alter the soil pH, improving nutrient availability. For example, citric acid released by the roots can help increase the availability of iron, calcium, and magnesium in the soil.

The Role of Soil Organic Matter

Organic matter in the soil, consisting of decaying plant and animal material, plays a crucial role in improving the soil's structure and fertility. It has the capacity to hold water and nutrients, creating a more favorable environment for root growth. Organic matter also promotes the activity of soil microorganisms, which decompose organic material and release nutrients that can be absorbed by plants. Humus, the stable form of organic matter, enhances soil structure, increases cation exchange capacity (CEC), and improves nutrient retention. Soils rich in organic matter have better water-holding capacity, greater nutrient availability, and more active microbial life, which all contribute to improved plant nutrition and growth.

The Rhizosphere and Microbial Interactions

The rhizosphere is the zone of soil around plant roots that is influenced by root activity. It is a highly dynamic environment where plant roots exude various substances, including sugars, amino acids, and organic acids, which serve as food sources for soil microorganisms. These microorganisms, in turn, can enhance nutrient availability to plants by breaking down organic matter, fixing nitrogen, or solubilizing minerals that are otherwise unavailable.

For example:

- **Nitrogen-fixing bacteria** (*e.g.*, Rhizobium in legumes) convert atmospheric nitrogen into a form that plants can use.
- **Phosphate-solubilizing bacteria** release phosphate from insoluble compounds in the soil, making it available to plants.
- **Mycorrhizal fungi** extend the root system and facilitate the uptake of nutrients, particularly phosphorus, by the plant.

Factors Affecting Nutrient Uptake

Several factors influence the efficiency of nutrient uptake from the soil:

- **Soil Fertility**: The availability of nutrients in the soil is a key determinant of plant growth. Fertile soils contain an adequate supply of essential macronutrients (nitrogen, phosphorus, potassium) and micronutrients (iron, zinc, copper) in forms that can be absorbed by plant roots. Regular soil testing and the appropriate application of fertilizers can help correct nutrient deficiencies and optimize plant growth.
- **Root System Development**: The development of a healthy, extensive root system is critical for maximizing nutrient uptake. Roots that penetrate deeply and spread widely can access a greater volume of soil, increasing the plant's ability to take up water and nutrients.
- **Soil Moisture**: The availability of water in the soil directly impacts nutrient uptake. Water helps dissolve nutrients in the soil and transports them to the roots. However, excess water can lead to nutrient leaching

and root oxygen deprivation, while insufficient water can slow down nutrient absorption, leading to deficiencies.

- **Soil Temperature**: Soil temperature affects the rate of nutrient uptake, as most biochemical processes in the plant (including nutrient absorption) occur more efficiently at moderate temperatures. Extremes in temperature (either too hot or too cold) can reduce the rate of nutrient uptake and slow down root growth.
- **Soil Aeration**: Oxygen is essential for root respiration and the active transport of nutrients. Poorly aerated soils, such as those that are waterlogged or compacted, restrict the oxygen supply to roots and hinder nutrient uptake.

Nutrient Deficiencies and Plant Growth

When plants are unable to obtain sufficient amounts of essential nutrients, they exhibit characteristic deficiency symptoms. These symptoms often first appear in the older leaves, as many nutrients are mobile and can be translocated from older to younger tissues. Nutrient deficiencies can severely impair plant growth, reduce yield, and make plants more susceptible to diseases. For example, nitrogen deficiency causes chlorosis (yellowing) of leaves, phosphorus deficiency results in stunted growth and purpling of leaves, and iron deficiency leads to interveinal chlorosis in younger leaves.

Transport of Nutrients and Water through the Plant

The transport of nutrients and water within a plant is a highly coordinated process, essential for plant growth, metabolism, and overall health. Plants rely on a complex vascular system to move water, minerals, and sugars between different parts of the plant. The main transport pathways in plants are through the xylem and phloem, two types of vascular tissues that carry out distinct but interconnected roles in fluid transport. Additionally, nutrient uptake occurs at the roots, while water is absorbed from the soil, and both substances are transported upward to the stems, leaves, and other growing parts of the plant. The process of nutrient and water movement is influenced by environmental factors, plant physiology, and the interaction between different transport mechanisms.

Water Transport Through the Plant

Water transport begins at the roots, where it is absorbed from the soil through the process of osmosis. Once in the root system, water moves into the xylem vessels, specialized tubes that carry water and dissolved minerals upward through the plant. The movement of water through the plant occurs through several mechanisms:

- **Root Pressure**: At the root, water is absorbed by the root cells and enters the xylem through a process known as osmosis. The accumulation of water creates a positive pressure in the root xylem, known as root pressure, which can help push water upward, particularly in small plants or during the night when transpiration is minimal.

- **Capillarity and Cohesion-Tension Theory**: The primary mechanism driving water movement through the plant is the cohesion-tension theory. As water evaporates from the leaves through transpiration, it creates a negative pressure or tension within the leaf. This tension is transmitted down through the xylem, pulling water upward from the roots to the leaves. The cohesion between water molecules (due to hydrogen bonding) ensures that the water column in the xylem remains intact, allowing the upward movement of water against gravity. Capillary action, which is the ability of water to rise in narrow tubes due to adhesive forces between water molecules and the walls of the xylem, further facilitates this process.
- **Transpiration Pull**: Transpiration, the loss of water vapor from the stomata in the leaves, creates a negative pressure that draws water from the soil through the roots and into the plant's vascular system. This process is known as the transpiration pull, and it is a continuous process that helps maintain the flow of water through the plant. The movement of water from the roots to the leaves helps transport nutrients and minerals dissolved in the soil solution, as well as maintaining turgor pressure in the plant cells.

Nutrient Transport Through the Plant

While water is primarily transported through the xylem, the movement of nutrients–such as minerals, ions, and other dissolved substances–relies on both active and passive transport mechanisms. Nutrients are absorbed from the soil by the roots and are subsequently transported throughout the plant to meet its metabolic needs.

- **Uptake of Nutrients by Roots**: The process of nutrient uptake begins when plant roots absorb water from the soil, carrying dissolved minerals into the root system. This absorption occurs primarily through root hairs, which increase the surface area available for nutrient uptake. Nutrients such as nitrogen (in the form of nitrate or ammonium), phosphorus (as phosphate), and potassium (as potassium ions) are taken up by the root cells via mechanisms such as active transport, facilitated diffusion, and ion exchange. In some cases, plant roots work with mycorrhizal fungi, which extend the root system and help in the absorption of nutrients, particularly phosphorus.
- **Transport of Nutrients via the Xylem**: Once absorbed by the roots, nutrients enter the xylem and are transported upwards through the plant in the flow of water. The xylem sap contains dissolved minerals, sugars, and other organic molecules that are essential for plant metabolism. The upward movement of nutrients follows the same physical forces that drive

water transport–mainly transpiration pull–which carries both water and dissolved nutrients from the roots to the leaves and other growing tissues.

- **Phloem Transport of Sugars and Other Organic Molecules**: While the xylem primarily transports water and minerals, phloem is responsible for transporting organic compounds, primarily sugars produced during photosynthesis. This process is known as translocation. Sugars, mainly sucrose, are transported from the source tissues (usually leaves, where photosynthesis occurs) to the sink tissues (such as roots, fruits, or growing shoots), where they are either used for energy, growth, or stored for later use. The phloem operates on a pressure-flow mechanism, where sugars are loaded into the phloem by active transport at the source, creating a high osmotic pressure. This pressure pushes the sugar solution through the phloem vessels toward the sinks, where it is unloaded and used or stored.

Integration of Water and Nutrient Transport

The transport of water and nutrients is not an isolated process but rather one that is tightly integrated and interdependent. Water movement through the plant helps transport essential nutrients from the roots to the leaves and other parts. Similarly, the movement of nutrients helps maintain the plant's overall homeostasis, ensuring that each cell has the necessary building blocks to carry out metabolic processes.

- **Water and Nutrient Absorption**: The uptake of water by the roots creates a soil-to-plant gradient that allows for the absorption of nutrients dissolved in the water. Nutrients are absorbed through both passive transport (*e.g.*, diffusion and facilitated diffusion) and active transport (*e.g.*, through proton pumps and nutrient-specific transport proteins). As water is absorbed, it carries these nutrients into the plant's vascular system.
- **Role of Transpiration**: Transpiration is a key driver in nutrient transport. The process not only creates the negative pressure required for the movement of water through the plant but also allows nutrients to be carried upward with the water. The evaporation of water from the stomata in the leaves creates a continuous flow of water and dissolved minerals, ensuring that the plant's nutritional needs are met. The loss of water through transpiration also helps regulate plant temperature and maintain the balance of essential ions.
- **Nutrient Redistribution**: In addition to transporting nutrients from the roots to the leaves, plants also redistribute nutrients within the plant. Nutrients that are more mobile (*e.g.*, nitrogen, phosphorus, potassium) can be moved from older tissues to newer, actively growing tissues when there is a deficiency. However, some nutrients, such as calcium and

boron, are less mobile and tend to accumulate in younger tissues. This redistribution ensures that the plant can maintain metabolic functions, even under nutrient stress conditions.

Regulation of Water and Nutrient Transport

The transport of both water and nutrients is tightly regulated by the plant to optimize resource use. Several mechanisms ensure that the plant's needs are met without overloading the system with excess water or nutrients.

- **Stomatal Regulation**: The opening and closing of stomata (small pores on the leaf surface) play a critical role in controlling water loss through transpiration. The stomata are regulated by guard cells, which open in response to light and carbon dioxide levels and close when the plant is under water stress. By controlling transpiration, the plant can regulate the amount of water lost to the atmosphere and prevent excessive dehydration.
- **Xylem Vessel Regulation**: The movement of water through the xylem is also regulated by the plant's need for water. When water supply is abundant, xylem vessels are fully active in transporting water to the upper parts of the plant. However, when water availability is limited, the plant may reduce water transport, or in extreme cases, cut off water supply to non-essential tissues to conserve water for vital processes like photosynthesis and root growth.
- **Phloem Loading and Unloading**: The movement of sugars and other organic compounds in the phloem is regulated by the plant's metabolic activity. Source tissues (leaves) load sugars into the phloem actively, creating a high osmotic pressure that drives the flow of sap. At the sink tissues (*e.g.*, roots, fruits, growing shoots), sugars are unloaded and converted into other forms like starch, which can be stored for later use or used directly in metabolic processes.

Chapter 3

Photosynthesis and Energy Metabolism

Photosynthesis

Photosynthesis is a vital biological process through which green plants, algae, and some bacteria convert light energy into chemical energy. This process enables these organisms to produce their own food by synthesizing glucose from carbon dioxide and water, using sunlight as the primary energy source. In addition to supporting the growth and energy needs of plants themselves, photosynthesis plays a crucial role in maintaining life on Earth by releasing oxygen into the atmosphere, which is essential for the respiration of most living organisms. Through this remarkable natural process, plants form the foundation of nearly all ecosystems, supporting diverse food webs and contributing to the global carbon cycle.

- ☆ **Energy.** The glucose molecules serve as fuel for cells: their chemical energy can be harvested through processes like cellular respiration and fermentation, which generate adenosine triphosphate–, a small, energy-carrying molecule–for the cell's immediate energy needs.
- ☆ **Fixed carbon.** Carbon from carbon dioxide–inorganic carbon–can be incorporated into organic molecules; this process is called carbon fixation, and the carbon in organic molecules is also known as fixed carbon. The carbon that's fixed and incorporated into sugars during photosynthesis can be used to build other types of organic molecules needed by cells.

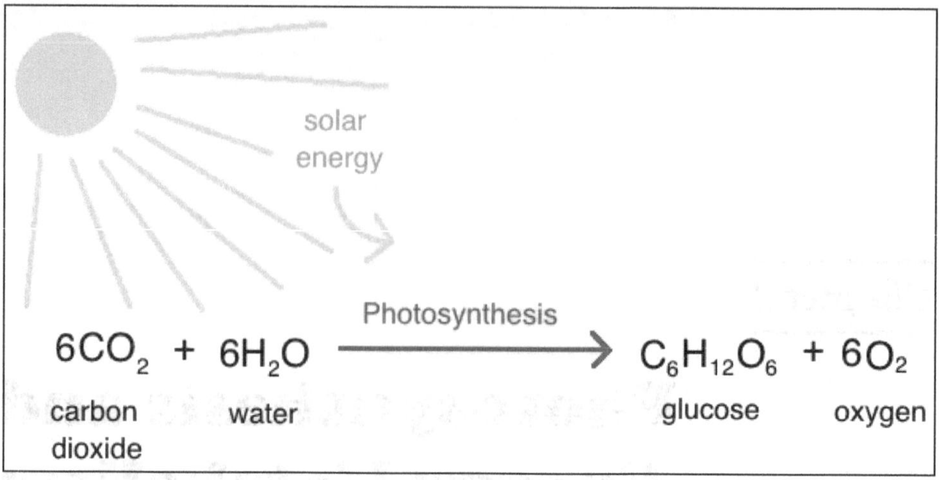

Figure 3.1

The Ecological Importance of Photosynthesis

Photosynthetic organisms, including plants, algae, and some bacteria, play a key ecological role. They introduce chemical energy and fixed carbon into ecosystems by using light to synthesize sugars. Since these organisms produce their own food–that is, fix their own carbon–using light energy, they are called photoautotrophs (literally, self-feeders that use light). Humans, and other organisms that can't convert carbon dioxide to organic compounds themselves, are called heterotrophs, meaning different-feeders. Heterotrophs must get fixed carbon by eating other organisms or their by-products. Animals, fungi, and many prokaryotes and protists are heterotrophs.

Besides introducing fixed carbon and energy into ecosystems, photosynthesis also affects the makeup of Earth's atmosphere. Most photosynthetic organisms generate oxygen gas as a byproduct, and the advent of photosynthesis–over billion years ago, in bacteria resembling modern cyanobacteria–forever changed life on Earth. These bacteria gradually released oxygen into Earth's oxygen-poor atmosphere, and the increase in oxygen concentration is thought to have influenced the evolution of aerobic life forms–organisms that use oxygen for cellular respiration. If it hadn't been for those ancient photosynthesizers, we, like many other species, wouldn't be here today! Photosynthetic organisms also remove large quantities of carbon dioxide from the atmosphere and use the carbon atoms to build organic molecules. Without Earth's abundance of plants and algae to continually suck up carbon dioxide, the gas would build up in the atmosphere. Although photosynthetic organisms remove some of the carbon dioxide produced by human activities, rising atmospheric levels are trapping heat and causing the climate to change. Many scientists believe that preserving forests and other expanses of vegetation is increasingly important to combat this rise in carbon dioxide levels.

Leaves are Sites of Photosynthesis

Plants are the most common autotrophs in terrestrial–land–ecosystems. All green plant tissues can photosynthesize, but in most plants, but the majority of photosynthesis usually takes place in the leaves. The cells in a middle layer of leaf tissue called the mesophyll are the primary site of photosynthesis.

Small pores called stomata–singular, stoma–are found on the surface of leaves in most plants, and they let carbon dioxide diffuse into the mesophyll layer and oxygen diffuse out.

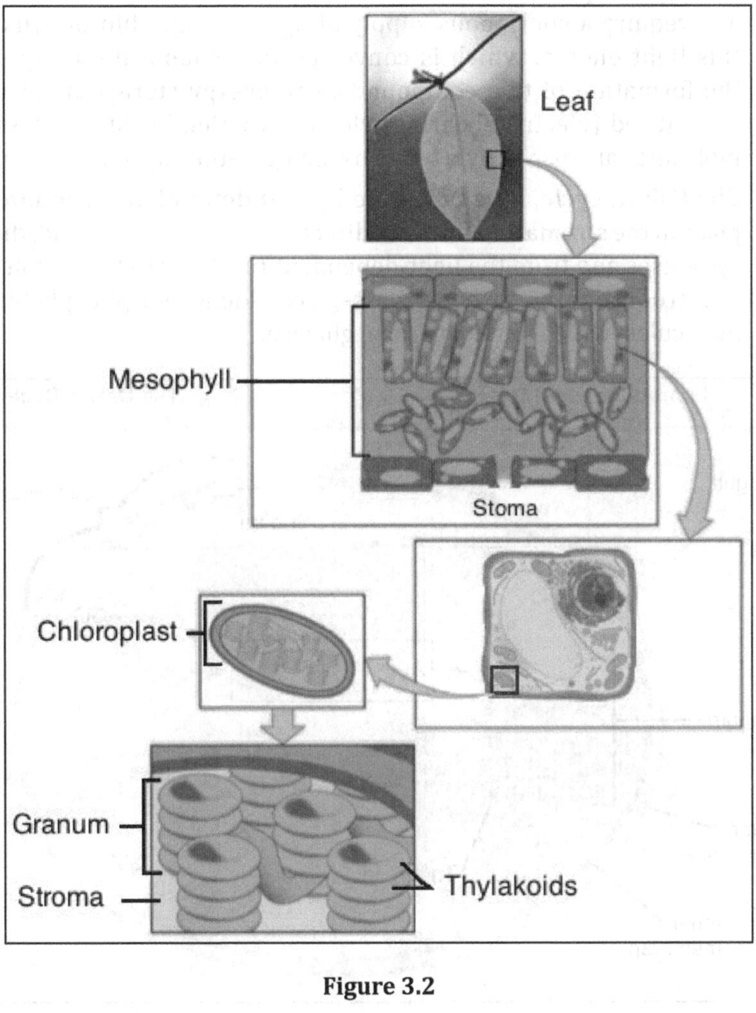

Figure 3.2

Each mesophyll cell contains organelles called chloroplasts, which are specialized to carry out the reactions of photosynthesis. Within each chloroplast, disc-like structures called thylakoids are arranged in piles like stacks of pancakes that are known as grana–singular, granum. The membrane of each thylakoid contains green-colored pigments called chlorophylls that absorb light. The fluid-filled space around the grana is called the stroma, and the space inside the thylakoid discs is known as the thylakoid space. Different chemical reactions occur in the different parts of the chloroplast.

The Light-Dependent Reactions and the Calvin Cycle

Photosynthesis in the leaves of plants involves many steps, but it can be divided into two stages: the light-dependent reactions and the Calvin cycle.

- The **light-dependent reactions** take place in the thylakoid membrane and require a continuous supply of light energy. Chlorophylls absorb this light energy, which is converted into chemical energy through the formation of two compounds, –an energy storage molecule–and –a reduced (electron-bearing) electron carrier. In this process, water molecules are also converted to oxygen gas–the oxygen we breathe!
- The **Calvin cycle**, also called the **light-independent reactions**, takes place in the stroma and does not directly require light. Instead, the Calvin cycle uses and from the light-dependent reactions to fix carbon dioxide and produce three-carbon sugars–glyceraldehyde-3-phosphate, or G3P, molecules–which join up to form glucose.

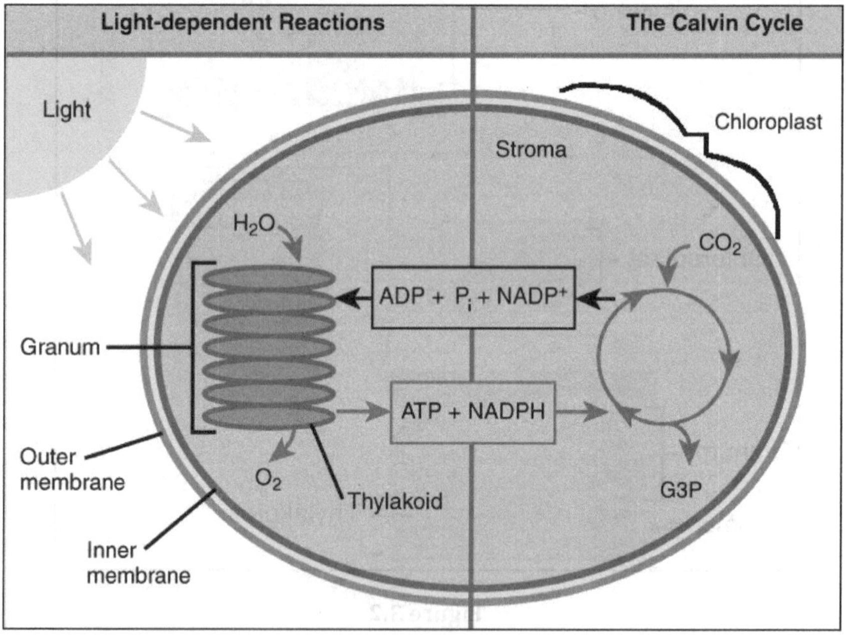

Figure 3.3

Overall, the light-dependent reactions capture light energy and store it temporarily in the chemical forms There is broken down to release energy, and donates its electrons to convert carbon dioxide molecules into sugars. In the end, the energy that started out as light winds up trapped in the bonds of the sugars.

Photosynthesis vs. Cellular Respiration

At the level of the overall reactions, photosynthesis and cellular respiration are near-opposite processes. They differ only in the form of energy absorbed or released, as shown in the diagram below.

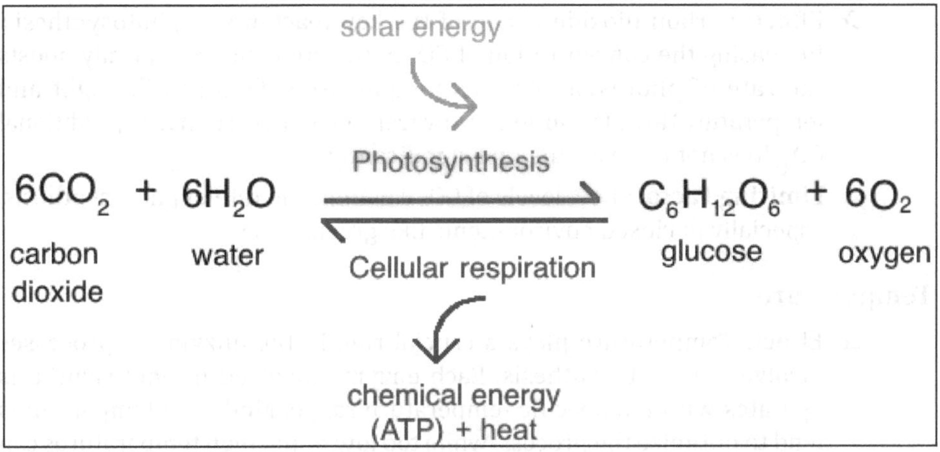

Figure 3.4

At the level of individual steps, photosynthesis isn't just cellular respiration run in reverse. Instead, as we'll see the rest of this section, photosynthesis takes place in its own unique series of steps. However, there are some notable similarities between photosynthesis and cellular respiration. For instance, photosynthesis and cellular respiration both involve a series of redox reactions (reactions involving electron transfers). In cellular respiration, electrons flow from glucose to oxygen, forming water and releasing energy. In photosynthesis, they go in the opposite direction, starting in water and winding up in glucose–an energy-requiring process powered by light. Like cellular respiration, photosynthesis also uses an electron transport chain to make a concentration gradient, which drives synthesis by chemiosmosis.

Factor Affecting Photosynthesis

Several factors influence the rate of photosynthesis, affecting how efficiently plants can produce energy and oxygen. These factors include environmental conditions, the availability of essential resources, and the internal health of the plant. Below are the key factors that affect photosynthesis:

Light Intensity

- ☆ **Effect:** Light provides the energy required for photosynthesis. As light intensity increases, the rate of photosynthesis generally increases, up to a certain point. Beyond that threshold, the plant's photosynthetic machinery becomes saturated, and further increases in light will not significantly enhance the rate of photosynthesis.

- **Limiting Factor:** If light intensity is too low, photosynthesis will be slower, and the plant may not produce enough energy for growth.

Carbon Dioxide Concentration (CO_2)

- **Effect:** Carbon dioxide is one of the key reactants in photosynthesis. Increasing the concentration of CO_2 in the atmosphere typically boosts the rate of photosynthesis, as long as other factors (like light and temperature) are also optimal. However, beyond a certain level, additional CO_2 does not increase the rate significantly.
- **Limiting Factor:** Low levels of CO_2 can limit the rate of photosynthesis, especially in closed environments like greenhouses.

Temperature

- **Effect:** Temperature plays a crucial role in the enzymatic processes involved in photosynthesis. Each enzyme involved in photosynthesis operates within a specific temperature range. Moderate temperatures tend to optimize the process, while too low or too high temperatures can slow down or even halt photosynthesis.
- **Limiting Factor:** Extremely low temperatures can cause enzymes to become inactive, while high temperatures may denature enzymes, both of which negatively impact photosynthesis.

Water Availability

- **Effect:** Water is an essential raw material for photosynthesis. It is split into oxygen, protons, and electrons during the light-dependent reactions. Adequate water is required for these reactions to proceed effectively.
- **Limiting Factor:** Water scarcity can cause stomata (pores on the leaf) to close, reducing CO_2 intake and slowing photosynthesis. Severe water stress can also damage plant cells and tissues.

Chlorophyll Concentration

- **Effect:** Chlorophyll is the pigment that absorbs light energy and converts it into chemical energy. Higher concentrations of chlorophyll increase the plant's ability to capture light and drive the photosynthetic process.
- **Limiting Factor:** A lack of chlorophyll (due to poor nutrition, disease, or genetic factors) can reduce the efficiency of photosynthesis.

Oxygen Concentration

- **Effect:** Oxygen is a by-product of photosynthesis and, at high concentrations, can affect the rate of photosynthesis through a process called photorespiration, where oxygen competes with carbon dioxide for the active site on the enzyme RuBisCO. Photorespiration can reduce the overall efficiency of photosynthesis.
- **Limiting Factor:** High oxygen levels can inhibit photosynthesis, especially in C3 plants.

Nutrient Availability

- **Effect:** Essential nutrients, especially nitrogen, phosphorus, potassium, and magnesium, are required for the synthesis of key compounds like chlorophyll and other enzymes involved in photosynthesis.
- **Limiting Factor:** Deficiencies in these nutrients can lead to reduced photosynthetic activity and poor plant health.

Soil pH

- **Effect:** Soil pH can affect nutrient availability and the health of the plant's root system. Extreme pH levels (either too acidic or too alkaline) can hinder nutrient uptake, affecting photosynthesis indirectly.
- **Limiting Factor:** Unfavorable soil pH may restrict the plant's ability to access essential minerals, thus slowing down the photosynthetic rate.

Environmental Stress (*e.g.*, Pollution, Pests, and Diseases)

- **Effect:** Environmental stress factors such as pollution, pests, and diseases can damage plant tissues, reduce chlorophyll content, and impede the function of photosynthetic enzymes, thereby reducing the efficiency of photosynthesis.
- **Limiting Factor:** Stress conditions can cause stomatal closure, reduce the plant's ability to absorb carbon dioxide, and damage cellular components essential for photosynthesis.

Plant Type and Genetic Factors

- **Effect:** Different plant species have varying photosynthetic efficiencies depending on their genetic makeup. For example, C4 and CAM plants are more efficient in certain environmental conditions, like high temperatures and low water availability, compared to C3 plants.
- **Limiting Factor:** The plant's inherent genetic traits determine its photosynthetic efficiency under various environmental conditions.

Table 3.1. Factors Affecting Photosynthesis

Factor	*Effect*	*Limiting Condition*
Light Intensity	Provides energy for photosynthesis; more light increases the rate.	Low light reduces photosynthetic activity.
CO_2 Concentration	More CO_2 generally increases photo-synthesis rate.	Low CO_2 limits the rate of photosynthesis.
Temperature	Affects enzyme activity in photosynthesis.	Too high or too low temperatures decrease efficiency.
Water Availability	Water is a key reactant in photosynthesis.	Drought conditions limit photosynthesis.
Chlorophyll Concentration	Higher chlorophyll enhances light absorption.	Low chlorophyll reduces photosynthetic efficiency.
Oxygen Concentration	High oxygen levels can inhibit photosynthesis (photorespiration).	High oxygen can reduce photosynthesis efficiency.
Nutrient Availability	Essential for chlorophyll production and enzyme function.	Nutrient deficiencies lower photosynthetic rates.
Soil pH	Affects nutrient uptake and root health.	Extreme pH disrupts nutrient absorption.
Environmental Stress	Stress factors like pollution and disease reduce efficiency.	Stress reduces photosynthetic capacity.
Plant Type	Different plant species have different efficiencies.	Genetic traits determine photosynthesis rates.

Chapter 4

Respiration in Plants

Introduction

Respiration in plants is a vital biological process that involves the release of energy from stored food, primarily in the form of glucose, to fuel various cellular activities. Unlike photosynthesis, which requires light and occurs mainly in the chloroplasts, respiration takes place continuously in all plant cells, both day and night. During respiration, plants take in oxygen and release carbon dioxide, breaking down glucose to produce energy in the form of adenosine triphosphate (ATP). This energy is essential for growth, maintenance, repair, and various metabolic functions. While respiration is similar in plants and animals, its role in the overall life cycle of plants is crucial for supporting their survival and reproduction.

Glycolysis and the Citric Acid Cycle

Glycolysis and the Citric Acid Cycle are two key stages of cellular respiration, the process by which cells generate energy from glucose. Both occur in the cytoplasm and mitochondria of plant cells, respectively, and play essential roles in breaking down glucose to release energy.

Glycolysis

Glycolysis is the first step in cellular respiration and takes place in the cytoplasm of the cell. It begins with one molecule of glucose (a six-carbon sugar) and, through a series of enzyme-catalyzed reactions, is broken down into two molecules of pyruvate (three carbons each). During this process, a small amount of energy is released and stored in the form of 2 ATP molecules and 2 NADH molecules (which carry electrons for later stages). Glycolysis does not require oxygen, making it an anaerobic process, but it is a critical first step regardless of whether oxygen is present for further aerobic respiration.

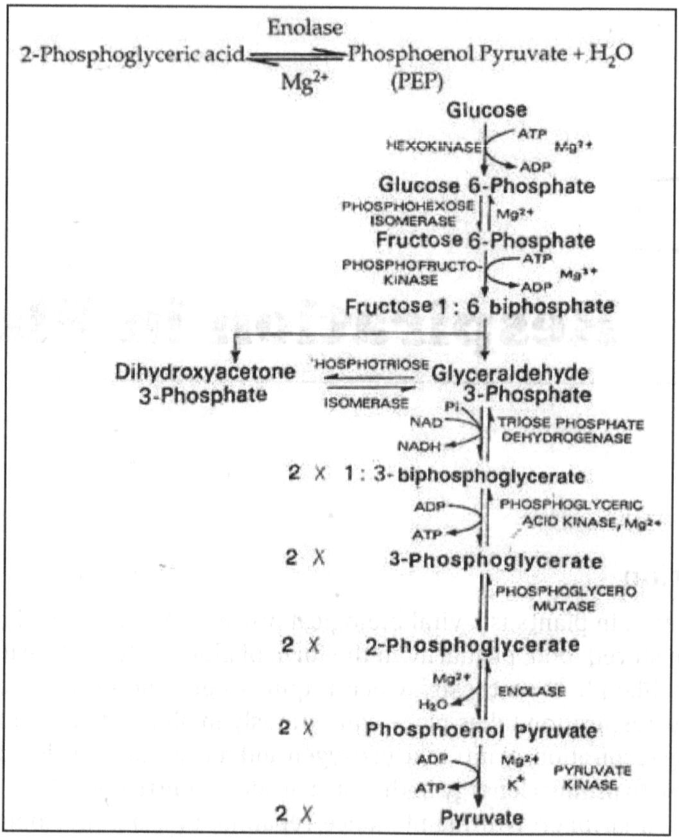

Figure 4.1. Schematic Representation of Glycolysis.

Citric Acid Cycle (Kreb's Cycle)

The citric acid cycle, also known as the Krebs cycle or tricarboxylic acid cycle, occurs in the mitochondria of plant cells, after glycolysis. The two pyruvate molecules produced in glycolysis are transported into the mitochondria, where each is converted into acetyl-CoA. Acetyl-CoA enters the citric acid cycle, combining with a four-carbon molecule called oxaloacetate to form citric acid (a six-carbon compound), from which the cycle gets its name. Over a series of reactions, citric acid is progressively broken down, releasing carbon dioxide and transferring high-energy electrons to carrier molecules, NADH and FADH. These molecules will later be used in the electron transport chain to produce ATP. For each glucose molecule that undergoes glycolysis, the citric acid cycle turns twice, generating a total of 2 ATP, 6 NADH, 2 FADH, and 4 CO_2 molecules.

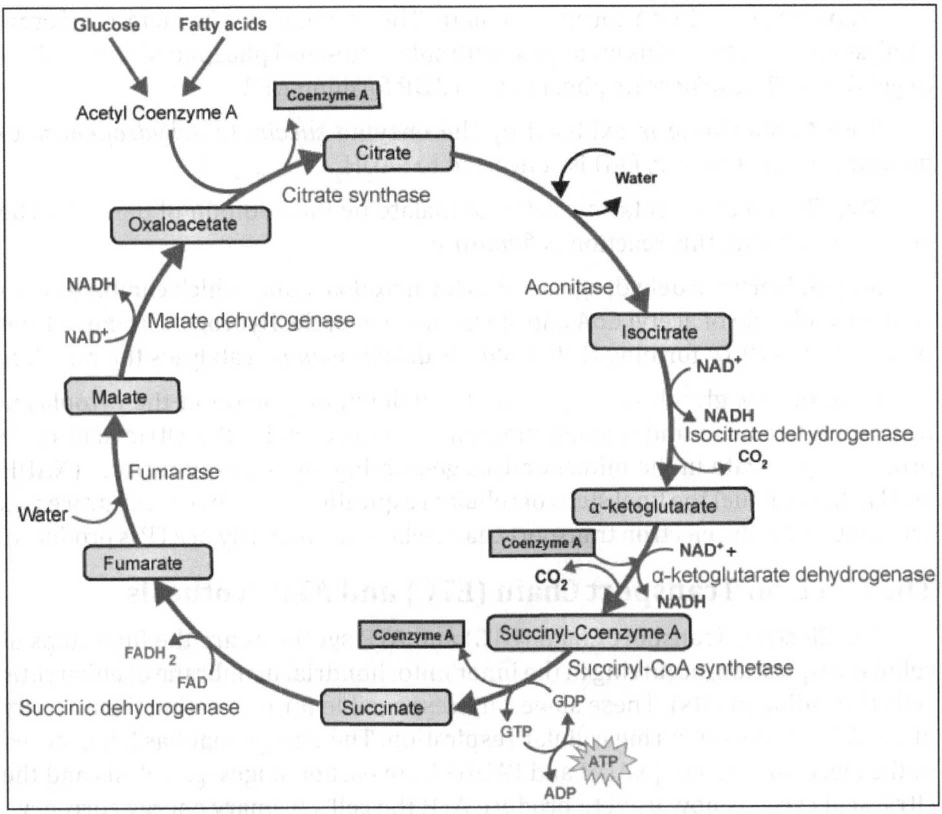

Figure 4.2. Schematic Representation of Citric Acid Cycle (Kreb's Cycle).

It is an eight-step process. Kreb's cycle or TCA cycle takes place in the matrix of mitochondria under aerobic condition.

Step 1: The first step is the condensation of acetyl CoA with 4-carbon compound oxaloacetate to form 6C citrate, coenzyme A is released. The reaction is catalysed by *citrate synthase.*

Step 2: Citrate is converted to its isomer, isocitrate. The enzyme *aconitase* catalyses this reaction.

Step 3: Isocitrate undergoes dehydrogenation and decarboxylation to form 5C 5p-ketoglutarate. A molecular form of CO_2 is released. *Isocitrate dehydrogenase* catalyses the reaction. It is an NAD^+ dependent enzyme. NAD^+ is converted to NADH.

Step 4: 5p-ketoglutarate undergoes oxidative decarboxylation to form succinyl CoA, a 4C compound. The reaction is catalyzed by the *5p-ketoglutarate dehydrogenase* enzyme complex. One molecule of CO_2 is released and NAD^+ is converted to NADH.

Step 5: Succinyl CoA forms succinate. The enzyme *succinyl CoA synthetase* catalyses the reaction. This is coupled with substrate-level phosphorylation of GDP to get GTP. GTP transfers its phosphate to ADP forming ATP.

Step 6: Succinate is oxidised by the enzyme *succinate dehydrogenase* to fumarate. In the process, FAD is converted to $FADH_2$.

Step 7: Fumarate gets converted to malate by the addition of one H_2O. The enzyme catalysing this reaction is *fumarase*.

Step 8: Malate is dehydrogenated to form oxaloacetate, which combines with another molecule of acetyl CoA and starts the new cycle. Hydrogens removed, get transferred to NAD^+ forming NADH. *Malate dehydrogenase* catalyses the reaction.

In summary, glycolysis begins the breakdown of glucose in the cytoplasm, producing pyruvate and a small amount of energy, while the citric acid cycle processes pyruvate in the mitochondria, generating high-energy carriers (NADH and FADH) that fuel the final steps of cellular respiration. Together, these processes set the stage for the electron transport chain, where the majority of ATP is produced.

The Electron Transport Chain (ETC) and ATP Synthesis

The Electron Transport Chain (ETC) and ATP synthesis are the final steps of cellular respiration, occurring in the inner mitochondrial membrane of eukaryotic cells (including plants). These stages are responsible for generating the majority of the ATP produced during cellular respiration. The energy that has been stored in the electron carriers (NADH and FADH) from earlier stages–glycolysis and the citric acid cycle–is now used to produce ATP, the cell's primary energy currency.

Electron Transport Chain (ETC)

The electron transport chain is a series of protein complexes and other molecules embedded in the inner mitochondrial membrane. Its primary role is to transfer electrons from high-energy carriers (NADH and FADH) to molecular oxygen (O_2), the final electron acceptor.

- **NADH and FADH Donate Electrons:** NADH and FADH produced earlier in the cytoplasm and mitochondria donate their high-energy electrons to the *etc.* NADH donates electrons to Complex I, while FADH donates electrons to Complex II.
- **Electron Transfer:** As electrons move through the chain of protein complexes (Complex I, II, III, and IV), they lose energy. This energy is used to actively pump protons (H^+ ions) from the mitochondrial matrix into the intermembrane space, creating a proton gradient (higher concentration of protons outside the inner membrane than inside).
- **Oxygen as the Final Electron Acceptor:** At the end of the chain, the low-energy electrons combine with protons (H^+) and oxygen (O_2) to form water (H_2O). Oxygen is essential in this process, as it is the final electron

acceptor, preventing the backup of electrons and allowing the chain to continue functioning.

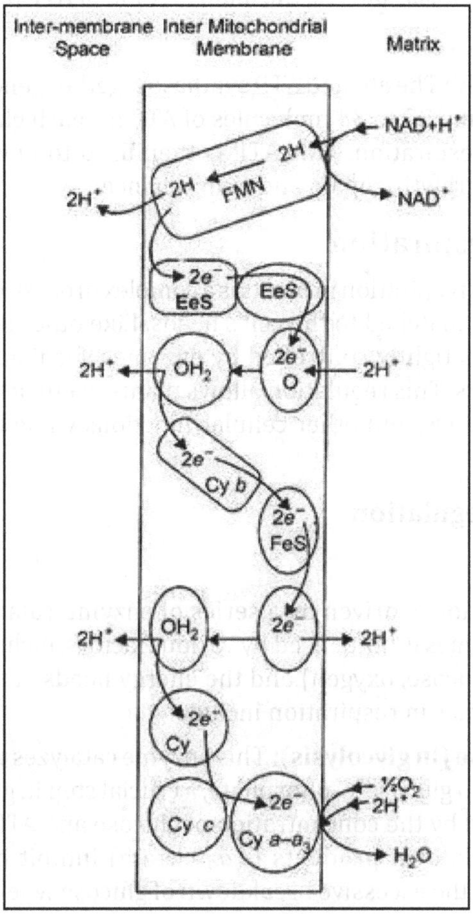

Figure 4.3: Electron Transport System (ETS).

ATP Synthesis (Chemiosmosis)

The proton gradient created by the *etc.* is key to driving the production of ATP. The concentration of protons is much higher in the intermembrane space than in the matrix, creating both a concentration gradient and an electrical gradient (proton motive force).

☆ **ATP Synthase:** The enzyme ATP synthase, which is also embedded in the inner mitochondrial membrane, acts like a turbine. Protons flow back into the mitochondrial matrix through ATP synthase due to the gradient, and this flow provides the energy needed to convert ADP (adenosine diphosphate) and inorganic phosphate (Pi) into ATP.

- **ATP Production:** As protons pass through ATP synthase, it rotates, catalyzing the phosphorylation of ADP to form ATP. This process is known as chemiosmosis, and it is the main method of ATP production in cellular respiration.

Overall ATP Yield: The *etc.* and ATP synthase together generate a large amount of ATP, typically around 32 to 34 molecules of ATP for each glucose molecule that undergoes cellular respiration. This ATP is then used to power various cellular activities, including growth, repair, and maintenance.

Regulation of Respiration

The regulation of respiration in plants is a complex process, ensuring that energy production is closely matched to the cell's needs. Like other metabolic processes, cellular respiration is tightly controlled by enzymes, feedback mechanisms, and environmental factors. This regulation allows plants to efficiently produce energy for growth, maintenance, and other cellular functions while preventing wasteful energy expenditure.

Mechanisms of Regulation

Enzyme Control

Cellular respiration is driven by a series of enzyme-catalyzed reactions. The activity of these enzymes is influenced by various factors, including the availability of substrates (*e.g.*, glucose, oxygen) and the energy needs of the cell. Some of the key regulatory enzymes in respiration include:

- **Hexokinase (in glycolysis):** This enzyme catalyzes the phosphorylation of glucose to glucose-6-phosphate, a crucial step in glycolysis. Its activity is regulated by the concentration of glucose and ATP. High levels of ATP or its breakdown products (*e.g.*, citrate) inhibit hexokinase activity, preventing the excessive breakdown of glucose when energy is plentiful.
- **Phosphofructokinase (PFK, in glycolysis):** PFK is a major regulatory enzyme in glycolysis and is controlled by feedback inhibition. It is activated by AMP (indicating low energy status) and inhibited by ATP and citrate (indicating high energy status).
- **Pyruvate Dehydrogenase (in the citric acid cycle):** This enzyme controls the conversion of pyruvate to acetyl-CoA, a critical step linking glycolysis and the citric acid cycle. It is inhibited by high levels of ATP and NADH, signaling that the cell has enough energy and does not need to continue **the cycle.**
- **Citrate Synthase (in the citric acid cycle):** This enzyme is involved in the first step of the citric acid cycle. It is regulated by feedback inhibition from ATP and NADH, which slows down the cycle when energy levels are high.

Feedback Inhibition

One of the most important regulatory mechanisms in respiration is feedback inhibition, where the end products of the respiration pathway (such as ATP and NADH) inhibit the activity of key enzymes earlier in the pathway. This helps to balance energy production with energy demand:

- When ATP levels are high, it signals that the cell has sufficient energy, leading to the inhibition of enzymes like phosphofructokinase (PFK) and pyruvate dehydrogenase, slowing down respiration.
- When ATP levels are low, ADP and AMP levels rise, activating enzymes like PFK to speed up glycolysis and the citric acid cycle, ensuring that more ATP is produced.

Oxygen Availability (Aerobic vs. Anaerobic Respiration)

The presence or absence of oxygen plays a critical role in regulating respiration in plants:

- **Aerobic Respiration:** When oxygen is available, plants undergo aerobic respiration, which is far more efficient, producing up to 36-38 ATP molecules per glucose molecule. Oxygen acts as the final electron acceptor in the electron transport chain, allowing the full oxidation of glucose.
- **Anaerobic Respiration:** In low-oxygen conditions, plants switch to anaerobic respiration, producing ATP less efficiently (only 2 ATP per glucose) via fermentation pathways such as alcoholic fermentation. This adaptation is common in roots or submerged plants where oxygen availability can be limited. However, anaerobic respiration produces byproducts like ethanol or lactic acid that can be toxic to the plant if accumulated.

Hormonal Regulation

Plant hormones also influence the rate of respiration, particularly in response to developmental or environmental changes:

- **Auxins:** These hormones can influence cellular respiration by promoting cell growth and division, which in turn increases the demand for energy and thus regulates the rate of respiration.
- **Cytokinins and Gibberellins**: Both hormones stimulate cell division and growth, leading to increased respiration to support the energy requirements of growing tissues.
- **Abscisic Acid (ABA):** ABA is typically involved in stress responses, such as during water shortage (drought stress), and can inhibit respiration to conserve energy.

Environmental Factors

Various environmental factors also regulate the rate of respiration in plants:

- **Temperature:** Respiration rate generally increases with temperature, as higher temperatures accelerate enzyme activity. However, if the temperature becomes too high, enzymes can denature, leading to a decrease in respiration efficiency. Conversely, lower temperatures slow down enzyme activity, reducing respiration.
- **Light:** While light does not directly affect respiration, it influences the balance between photosynthesis and respiration. During the day, plants photosynthesize and produce glucose, which may be used in respiration. At night, when photosynthesis stops, respiration continues to provide the plant with ATP for growth and other metabolic functions.

The regulation of respiration in plants ensures that energy production is balanced with the cell's needs and environmental conditions. It is controlled by enzyme activity, feedback inhibition, oxygen availability, hormones, and environmental factors. This finely tuned regulation allows plants to adapt to fluctuating conditions, conserve energy, and optimize their metabolic efficiency for growth, maintenance, and reproduction.

Chapter 5

Plant Growth and Development

Introduction

Plant growth and development are complex processes that enable plants to survive, reproduce, and adapt to their environments. Growth refers to the increase in size and mass of a plant, driven by cell division, elongation, and differentiation. Development encompasses the entire life cycle of a plant, from germination through to maturity and reproduction, including the formation of specialized structures like roots, stems, leaves, and flowers. The process of plant growth and development is regulated by a combination of genetic factors and environmental influences, with hormones playing a crucial role in coordinating various physiological responses. Key hormones, such as auxins, cytokinins, gibberellins, and abscisic acid, regulate processes like cell division, elongation, flowering, and responses to stress. Environmental factors such as light, temperature, water, and nutrient availability also influence plant growth and development. For example, light affects the process of photosynthesis, while temperature impacts metabolic rates and developmental timing. In understanding plant growth and development, scientists aim to decipher the intricate pathways that allow plants to respond to their surroundings, ensuring optimal growth and reproduction. This knowledge has vast applications in agriculture, horticulture, and environmental science, where improving plant growth can lead to better crop yields, sustainable farming practices, and enhanced biodiversity conservation.

Plant Growth: Cell Division, Expansion and Differentiation

Plant growth is a highly intricate and tightly regulated process that results from the coordinated actions of cell division, cell expansion, and cell differentiation. These three fundamental mechanisms work in concert to drive the increase in size and complexity of the plant, enabling it to acquire specialized functions, form diverse tissues, and develop organs essential for its survival, reproduction, and adaptation to the environment. Cell division generates the initial pool of new cells, which then expand to promote growth in size. As these cells enlarge, they undergo differentiation, acquiring distinct structures and functions that contribute to the formation of specialized tissues, such as vascular, dermal, and ground tissues. This dynamic interplay of processes allows plants to not only grow in size but also to form intricate networks of organs–such as roots, stems, leaves, and flowers–that are crucial for resource acquisition, structural support, reproduction, and environmental interaction. Together, these processes ensure that plants not only grow but also thrive and adapt in a constantly changing environment.

1. Cell Division

Cell division is the first step in plant growth and occurs primarily in the meristems, which are specialized regions of undifferentiated cells capable of continuous division. Meristems are found in two main areas: apical meristems (at the tips of roots and shoots) and lateral meristems (in the cambium, responsible for secondary growth in woody plants).

- ☆ **Mitosis**: Mitosis is a fundamental process of cell division in plant cells, where a single parent cell divides to produce two genetically identical daughter cells. This highly regulated process is essential for the growth, development, and maintenance of the plant, as it generates new cells to replace old or damaged ones and supports the formation of tissues and organs. During mitosis, the plant cell's DNA is precisely replicated and evenly distributed, ensuring that each daughter cell receives an exact copy of the genetic material. This accuracy is crucial for maintaining genetic stability and promoting proper functioning in the plant. Mitosis not only contributes to primary growth–such as the elongation of roots and shoots–but also plays a key role in the regeneration and repair of plant tissues. By enabling continuous cell production, mitosis ensures the plant can grow, adapt to environmental changes, and ultimately sustain its life cycle.
- ☆ **Cell Cycle**: The cell cycle is a highly organized and tightly regulated series of events that guide a cell through its growth, DNA replication, and division processes. It consists of two major phases: interphase and the mitotic phase (M phase). Interphase, which makes up the majority of the cell cycle, is subdivided into three distinct phases: G1 (Gap 1), S (Synthesis), and G2 (Gap 2). During G1 phase, the cell undergoes rapid growth, synthesizing proteins and organelles necessary for its function.

It also monitors internal and external conditions to ensure they are favorable for division. In the S phase, the cell's DNA is precisely replicated, ensuring that two identical copies of the genetic material are available for the upcoming division. The G2 phase follows, during which the cell continues to grow, producing additional proteins and organelles while checking the integrity of its DNA to ensure it is ready for mitosis. Once interphase is complete, the cell enters the M phase (mitotic phase), where the actual division process takes place. Mitosis is responsible for the separation of the replicated chromosomes into two identical sets, followed by cytokinesis, which divides the cytoplasm, forming two distinct daughter cells. Each daughter cell inherits a complete set of genetic instructions, ensuring it is genetically identical to the parent cell. Following division, the newly formed daughter cells enter G1 phase of their own cycle, where they begin to grow, carry out their specific functions, and, in some cases, differentiate into specialized cell types. This progression from G1 through the stages of interphase and mitosis is crucial for maintaining the health and integrity of the plant, facilitating continuous growth, tissue development, and response to environmental cues. The precise control of the cell cycle ensures that plant cells divide at the appropriate time, with accurate DNA replication, to support overall plant growth and adaptation.

- ✰ **Apical Meristem**: The apical meristem, found at the tips of the roots and shoots, is responsible for primary growth. Cells in the apical meristem continuously divide, allowing the plant to increase in length. These cells then undergo further processes of expansion and differentiation.
- ✰ **Lateral Meristem (Cambium)**: In woody plants, lateral meristems like the vascular cambium and cork cambium contribute to secondary growth (growth in thickness). The vascular cambium produces new xylem (wood) and phloem (bark), while the cork cambium generates protective bark tissue.

2. Cell Expansion

Cell expansion is the process by which newly formed cells increase in size. This is a crucial step for plant growth, as it contributes significantly to the enlargement of plant organs (roots, stems, leaves, flowers).

- ✰ **Turgor Pressure**: The expansion of plant cells is driven by turgor pressure, which is the pressure exerted by the cell wall against the internal water content (vacuole). As water enters the vacuole, it creates internal pressure, forcing the cell to expand.
- ✰ **Cell Wall Remodeling**: For cells to expand, the structure of the cell wall must be modified. The cell wall is primarily composed of cellulose, hemicellulose, and pectin. During expansion, enzymes such as expansins

loosen the bonds between cellulose fibers, allowing the cell wall to stretch and accommodate increased cell volume.

- **Direction of Expansion**: The direction of cell expansion is influenced by the orientation of cellulose microfibrils in the cell wall. In most plant cells, the microfibrils are arranged in a way that limits expansion in one direction (typically radially) and allows elongation along the axis of growth (axially). For instance, in the elongation of roots, cells mainly expand longitudinally, whereas in leaves, cells expand in both directions.
- **Factors Affecting Expansion**: Growth regulators like auxins play a significant role in promoting cell expansion. In certain tissues, auxins accumulate on the side of the organ experiencing more growth, creating a differential growth pattern (*e.g.*, phototropism, gravitropism).

3. Cell Differentiation

After cells divide and expand, they undergo differentiation, a process where unspecialized cells acquire specific structures and functions, allowing them to perform specialized tasks within the plant. Differentiation leads to the formation of various cell types that make up the plant's tissues and organs.

- **Totipotency and Pluripotency**: Plant cells retain a high degree of totipotency, meaning that a single plant cell can regenerate into a whole new plant under appropriate conditions. However, as cells undergo differentiation, they become pluripotent, meaning they acquire specific fates and become specialized for particular roles in the plant.
- **Tissue Differentiation**: Differentiation occurs in different types of tissues:
    - **Dermal Tissue**: This outer tissue includes the epidermis and specialized structures like trichomes (hair-like structures), guard cells (that regulate stomatal openings), and root hairs (for water and nutrient absorption).
    - **Vascular Tissue**: Vascular tissues include the xylem, responsible for water and nutrient transport, and the phloem, which transports sugars and organic compounds. Xylem cells differentiate into tracheids or vessel elements, while phloem cells differentiate into sieve tube elements and companion cells.
    - **Ground Tissue**: This tissue includes parenchyma, collenchyma, and sclerenchyma. Parenchyma cells are primarily involved in photosynthesis and storage, collenchyma cells provide flexible support, and sclerenchyma cells provide rigid support with thick, lignified walls.
    - **Meristematic Tissue**: Some cells in meristematic tissue retain the ability to continue dividing, while others differentiate into specialized

tissues. The role of meristematic tissue is central to both primary and secondary growth.

- **Hormonal Regulation of Differentiation**: Plant hormones, or phytohormones, such as auxins, cytokinins, gibberellins, and abscisic acid, coordinate the differentiation process. Auxins, for example, promote the formation of root primordia, while cytokinins encourage cell division and the formation of shoot tissues. Gibberellins help in stem elongation, and abscisic acid is involved in stress responses.
- **Developmental Pathways**: The process of differentiation is influenced by both gene expression and environmental factors. Gene regulatory networks determine the fate of cells, with different combinations of transcription factors, signaling pathways, and epigenetic modifications guiding whether a cell becomes part of the root, stem, leaf, or flower.

Interaction of Cell Division, Expansion, and Differentiation

The interaction of cell division, expansion, and differentiation is essential for coordinated plant growth and development. Cell division generates the initial pool of new cells, which are crucial for increasing plant size and creating the diverse tissues required for the plant's various functions. Once cells are produced through mitosis, they undergo cell expansion, where they increase in size, driven by turgor pressure and cell wall loosening. This expansion allows plant organs such as roots, stems, and leaves to elongate and take shape. After expansion, the cells begin to differentiate, acquiring specialized functions that enable them to form the various tissue types–such as xylem, phloem, and epidermis–that make up the plant's organs. The timing and coordination of these processes are tightly regulated by hormonal signals and environmental cues, ensuring that cells divide, expand, and differentiate in the correct sequence and location. This dynamic interplay allows plants to grow in both size and complexity, producing tissues and organs that enable them to capture light, absorb nutrients, transport water, and ultimately reproduce. The seamless coordination of cell division, expansion, and differentiation ensures that the plant can adapt to its environment and successfully complete its life cycle. Plant growth is not the result of any one of these processes in isolation. Instead, they are highly interdependent:

- Cell division creates new cells.
- Cell expansion increases the size of the plant.
- Cell differentiation ensures that these new, expanding cells take on specific roles in the plant's tissues and organs.

Together, these processes allow plants to grow, develop, and adapt to their environment. For example, a plant's roots grow longer by division of cells in the root meristem, followed by expansion of these cells, and eventually differentiation into specialized tissues like epidermal cells (for nutrient absorption) and xylem (for water transport). Moreover, environmental factors (light, gravity, temperature, *etc.*)

and internal signals (hormones, gene expression) further modulate these processes to optimize plant growth and development in response to changing conditions. The mechanisms of cell division, expansion, and differentiation are fundamental to plant growth and development. By coordinating these processes, plants can produce new tissues, organs, and systems that are essential for their survival and reproduction. Understanding how these processes work together provides valuable insights into plant biology and offers opportunities to improve agricultural practices, enhance crop yields, and create more resilient plant species.

The Role of Plant Hormones in Growth Regulation

Plant hormones, also known as phytohormones, are crucial chemical messengers that regulate nearly every aspect of plant growth, development, and response to environmental stimuli. These hormones are produced in small quantities but have profound effects on processes such as cell division, elongation, differentiation, and response to stress. The five major classes of plant hormones–auxins, cytokinins, gibberellins, abscisic acid, and ethylene–each play unique roles in coordinating plant growth and adapting the plant to changing conditions.

Plant hormones are essential regulators that govern plant growth, development, and responses to environmental cues. By coordinating processes such as cell division, elongation, and differentiation, hormones ensure that plants grow in a way that is both efficient and adaptive. Their roles in stress responses, reproduction, and environmental interactions highlight their importance in enabling plants to survive and thrive in diverse conditions. Understanding the mechanisms behind plant hormone action not only deepens our understanding of plant biology but also has practical applications in agriculture, horticulture, and environmental management, where manipulating hormone pathways can lead to enhanced crop productivity, stress resistance, and more sustainable farming practices.

Auxins

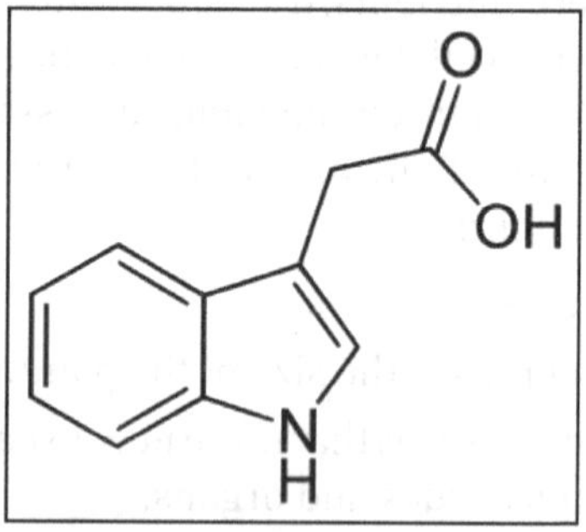

Auxins are a class of plant hormones that play a crucial role in regulating a variety of growth and developmental processes in plants, such as cell elongation, root initiation, apical dominance, phototropism, gravitropism, and flowering. They are one of the most important and well-studied plant hormones due to their wide-ranging effects on plant growth and development.

Structure of Auxins

The most common and well-known auxin is indole-3-acetic acid (IAA), a simple aromatic compound. IAA is a naturally occurring auxin that has been widely studied in plant biology. The structure of IAA consists of:

- **Indole Ring**: The indole ring is a six-membered aromatic ring containing nitrogen, derived from the amino acid tryptophan. This structure is responsible for the hormone's interaction with plant cell receptors.
- **Acetic Acid Group (–COOH)**: The acetic acid group attached to the indole ring allows IAA to function as a growth regulator, as it participates in its movement within the plant and its interaction with receptors on target cells.

In addition to IAA, there are other naturally occurring auxins in plants, such as indole-3-butyric acid (IBA), 2,4-dichlorophenoxyacetic acid (2,4-D), and synthetic auxins like naphthaleneacetic acid (NAA), which are used in agricultural practices.

Functions of Auxins in Plant Growth

Auxins are involved in a wide range of growth-promoting and developmental processes in plants. Their effects depend on the concentration, location, and interaction with other hormones. Some of the key functions of auxins include:

- **Cell elongation:** Auxins are primarily known for promoting cell elongation, particularly in the shoot. They stimulate the acidification of the cell wall by increasing the activity of proton pumps that transport protons (H^+) into the cell wall. The resulting acidic environment loosens the cellulose fibers in the wall, making it more flexible and allowing the cells to expand as they absorb water. this process is vital for the growth of stems and leaves.
- **Apical dominance:** Auxins play a central role in maintaining apical dominance, which is the phenomenon where the main central stem (or "apical bud") suppresses the growth of lateral (side) buds. The apical bud produces Auxins, which are transported downward into the stem and act to inhibit the growth of axillary buds. When the apical bud is removed (*e.g.*, by pruning), the suppression is lifted, and lateral buds begin to grow more vigorously.
- **Root development and initiation:** Auxins are crucial in promoting the formation of roots, particularly in the process of adventitious rooting, where roots form from non-root tissues like stems or leaves. High concentrations of auxins, particularly IBA, can stimulate root formation, which is why auxins are commonly used in rooting hormones to help propagate plants from cuttings.
- **Phototropism and gravitropism:** Auxins are responsible for the directional growth responses of plants to light (phototropism) and gravity

(gravitropism). In phototropism, auxins accumulate on the shaded side of a plant, causing those cells to elongate more than the cells on the light-exposed side, leading the plant to bend toward the light source. Similarly, in gravitropism, auxins accumulate on the lower side of roots and stems in response to gravity, directing the root to grow downward (positive gravitropism) and the stem to grow upward (negative gravitropism).

- ☆ **Flower and fruit development:** Auxins also play a role in the development of flowers and fruit. in some plants, auxins are involved in the formation of flower buds and the promotion of fruit set after pollination. The presence of auxins in the developing fruit can prevent abscission (the shedding of flowers or fruits), allowing the fruit to mature.
- ☆ **Xylem and phloem differentiation:** Auxins influence the differentiation of vascular tissues, such as the formation of xylem and phloem. They promote the differentiation of vascular cambium cells into secondary xylem (wood) and phloem, which are essential for long-distance transport of water, nutrients, and sugars.

Mechanisms of Action of Auxins

Auxins exert their effects through both local and systemic signaling mechanisms. Here are the key steps in their action:

Transport of Auxins: Auxins are produced primarily in the apical meristems (tip of the shoot and root), but they can be transported throughout the plant via two main pathways:

- ☆ **Polar Transport**: Auxins are actively transported from cell to cell in a unidirectional (polar) manner, primarily through the vascular tissue. The direction of transport is regulated by specific transport proteins known as **PIN proteins**, which help auxins move from cell to cell. This directional movement allows auxins to concentrate in certain areas of the plant, such as the shaded side of a stem, promoting differential growth responses like phototropism.
- ☆ **Non-Polar Transport**: Auxins can also be passively transported across cell membranes, often through diffusion or carrier-mediated transport, although this process is less specific than polar transport.

Auxin Receptors and Signaling: Once auxins reach their target cells, they bind to specific receptors called TIR1 (Transport Inhibitor Response 1) proteins, which are part of a larger receptor complex. This interaction triggers a signaling cascade that leads to the activation or suppression of gene expression. The major effects of auxins are mediated through the Aux/IAA proteins, which regulate the expression of auxin-responsive genes.

Gene Regulation: When auxins bind to their receptors, they promote the degradation of Aux/IAA repressors, which normally inhibit the activity of ARF

(Auxin Response Factor) transcription factors. Once the repression is lifted, ARFs activate the expression of genes involved in various growth processes, such as cell wall loosening, cell division, and differentiation. This molecular cascade helps coordinate the various physiological effects of auxins in plants.

Role of Auxins in Plant Growth and Development

Auxins are indispensable for the overall growth and development of plants. Some of their primary roles in plant growth include:

- ☆ **Stem Elongation**: Auxins promote elongation of cells in the stem, contributing to overall plant height.
- ☆ **Root Formation**: Auxins are involved in the initiation and growth of new roots, particularly in tissue cultures or cuttings.
- ☆ **Organ Formation**: Auxins regulate the formation of leaves, flowers, and lateral branches by influencing cell division and differentiation.
- ☆ **Tropisms**: Auxins mediate plant responses to light (phototropism) and gravity (gravitropism), allowing plants to optimize their growth conditions.

Applications of Auxins

Auxins, particularly synthetic forms such as 2,4-D and NAA, are widely used in agriculture and horticulture:

- ☆ **Rooting Hormones**: Auxins are often applied to plant cuttings to stimulate root formation, making it easier to propagate new plants.
- ☆ **Weed Control**: Certain synthetic auxins, like 2,4-D, are used as herbicides to control broadleaf weeds by disrupting their growth patterns.
- ☆ **Fruit Set and Growth**: Auxins are applied to increase fruit set and prevent premature fruit drop in some crops, like tomatoes and apples.

Auxins are vital plant hormones that play a central role in regulating plant growth and development. Their ability to promote cell elongation, influence root and shoot development, and guide plant responses to light and gravity makes them indispensable for a plant's survival and adaptability. The precise control of auxin distribution and action allows plants to optimize their growth in response to internal and environmental signals, contributing to their ability to grow, reproduce, and thrive.

Cytokinins

Cytokinins are a class of plant hormones that play a crucial role in regulating various aspects of plant growth and development, particularly in cell division and differentiation. They are involved in processes such as shoot initiation, leaf senescence, apical dominance, and the regulation of root growth. Cytokinins are essential hormones in plants, coordinating a wide range of physiological processes to ensure that plants grow, develop, and adapt effectively to their environment. Their intricate interactions with other hormones like auxins, gibberellins, and abscisic acid help fine-tune plant responses to internal and external cues.

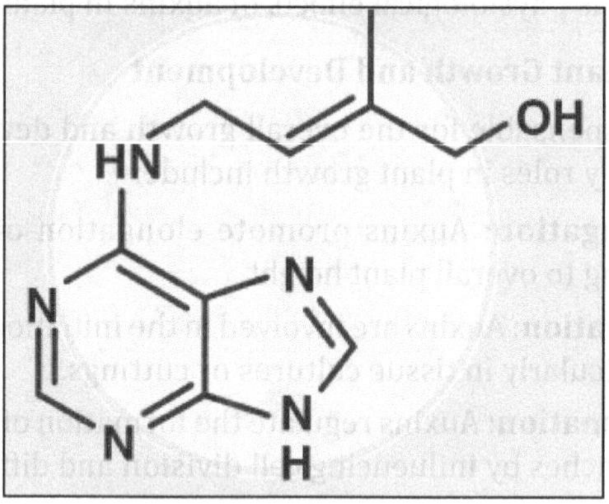

Cell Division and Cytokinesis

- ☆ **Promotion of Cell Division:** Cytokinins are directly involved in promoting cell division (cytokinesis) in plant cells. They activate the cell cycle, particularly during the G1 phase, where they push the cell toward entering the S phase of the cell cycle, where DNA replication occurs. This helps facilitate the formation of new cells, which is essential for plant growth and tissue formation.
- ☆ **Interaction with Other Hormones:** Cytokinins often work in balance with other hormones, like auxins. While auxins typically promote cell elongation and root development, cytokinins primarily regulate cell division and can influence the formation of shoot tissues. The interaction between cytokinin and auxin concentrations determines the type of growth a plant exhibits, such as root versus shoot development.

Shoot and Root Growth Regulation

- ☆ **Shoot Formation:** Cytokinins promote the development of shoots and buds. They are involved in the formation of lateral shoots and the outgrowth of axillary buds. High concentrations of cytokinins in the shoot system stimulate the growth of new shoots, while low levels allow the apical bud (tip of the shoot) to dominate. This is known as apical dominance.
- ☆ **Root Growth:** While cytokinins promote shoot growth, they inhibit root growth when present in high concentrations. In contrast, low levels of cytokinins combined with higher levels of auxins promote root development. This interplay between cytokinins and auxins is essential for coordinating the development of both roots and shoots.

Leaf Senescence and Longevity

- ☆ **Delay in Leaf Senescence:** Cytokinins play a role in delaying leaf senescence (aging) by promoting protein synthesis and maintaining chloroplast function. By delaying the breakdown of chlorophyll and other cell components, cytokinins help extend the life of leaves and maintain photosynthetic capacity. This is especially important during periods of stress when the plant needs to optimize resource use.
- ☆ **Regulation of Nutrient Transport:** Cytokinins help regulate the movement of nutrients between the leaves, roots, and other parts of the plant, which is critical for maintaining leaf health and function.

Vascular Tissue Development

- ☆ **Xylem and Phloem Formation:** Cytokinins influence the formation of vascular tissues like xylem (water transport) and phloem (nutrient and sugar transport). Cytokinins can promote the differentiation of vascular tissue from undifferentiated cells and help organize the structure of the vascular system in plant stems and leaves. This is crucial for efficient water and nutrient transport, especially in growing plant tissues.
- ☆ **Vascular Cambium Activity:** In woody plants, cytokinins play a role in the activity of the vascular cambium, which produces secondary xylem (wood) and phloem. This contributes to the thickening of stems and growth in girth.

Organ Development and Morphogenesis

- ☆ **Regulation of Organogenesis:** Cytokinins are essential for organ formation, including the initiation and development of both roots and shoots. They influence the growth of meristematic tissues, which are responsible for producing new organs in plants.
- ☆ **Callus Formation:** Cytokinins are commonly used in tissue culture to induce callus formation (undifferentiated plant tissue). When applied in specific concentrations, cytokinins can stimulate the formation of undifferentiated tissue, which can later differentiate into shoots or roots, depending on the presence of other hormones like auxins.

Stress Responses

- ☆ **Stress Tolerance:** Cytokinins are involved in plant responses to environmental stress, including drought, salinity, and heat. They help mitigate the effects of stress by regulating processes like water retention, stomatal regulation, and the synthesis of stress-related proteins. Cytokinins can also help improve the plant's ability to recover from stress by promoting cell repair and growth after damage.

- ☆ **Antioxidant Production:** Cytokinins can activate antioxidant pathways, which helps the plant combat oxidative stress and minimize damage from reactive oxygen species (ROS) during stressful conditions.

Cytokinin Signaling Pathway

- ☆ **Receptor-Mediated Activation:** Cytokinins bind to specific receptors on the cell surface (known as histidine kinases), initiating a signal transduction pathway that activates downstream responses in the cell. This signaling involves several proteins and pathways, including the two-component system (histidine kinase and response regulator proteins) that helps modulate gene expression.
- ☆ **Regulation of Gene Expression:** The cytokinin signaling pathway activates or represses genes involved in cell division, differentiation, and stress responses. This gene expression is coordinated with other signaling pathways to ensure the proper development of tissues and organs.

Role in Seed Germination

- ☆ **Promotion of Germination:** Cytokinins can play a role in seed germination by promoting cell division and the breakdown of stored food in seeds. In some cases, cytokinins are involved in the transition from dormant seed states to active growth, particularly in the early stages of seedling development.
- ☆ **Embryonic Development:** During early development, cytokinins are involved in regulating the growth of the embryo within the seed, influencing how tissues differentiate and develop into the mature plant.

Cytokinin Biosynthesis and Transport

- ☆ **Biosynthesis:** Cytokinins are synthesized in plant tissues, particularly in the roots, where they are produced by the enzyme isopentenyl transferase (IPT). These hormones are then transported throughout the plant via the xylem to other parts of the plant, such as the shoots and leaves.
- ☆ **Transport:** Cytokinins are primarily transported through the plant via xylem vessels, but they can also move to a lesser extent through the phloem. Their movement is regulated by factors such as local needs and environmental signals.

Roles of Cytokinins in Plant Growth and Development

- ☆ Promotion of cell division and cytokinesis, crucial for growth and tissue formation.
- ☆ Regulation of shoot and root development, balancing shoot growth and root inhibition.

☆ Delay of leaf senescence, enhancing the longevity of leaves for photosynthesis.

☆ Influence on vascular tissue formation for efficient water and nutrient transport.

☆ Regulation of organ formation and morphogenesis, such as root and shoot initiation.

☆ Improvement of stress tolerance, including responses to drought and other environmental challenges.

☆ Promotion of seed germination and embryonic development.

Gibberellins

Gibberellins (GAs) are a group of plant hormones that play a central role in promoting a wide range of growth and developmental processes, including seed germination, stem elongation, flowering, and fruit development. Gibberellins are among the most important growth regulators in plants and have diverse effects depending on the plant species and developmental stage. Gibberellins are essential hormones in plants that regulate a wide range of growth and developmental processes. From promoting seed germination and stem elongation to influencing flowering and fruit development, gibberellins have a profound impact on plant morphology and physiology. In agriculture, gibberellins are widely used to enhance crop growth, fruit size, and even to overcome dormancy in certain seeds.

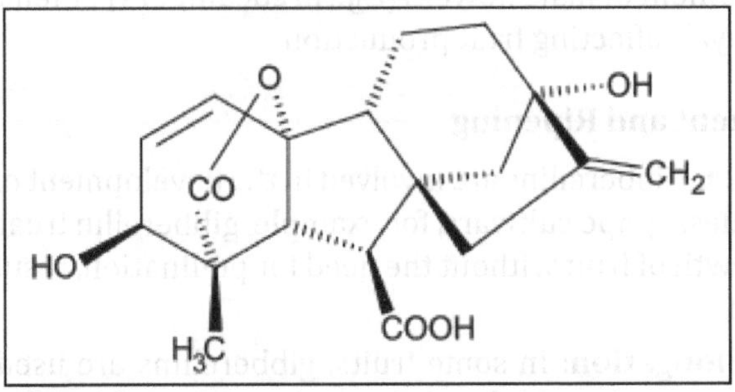

Chemical Structure of the Plant Hormone Gibberellic Acid.

Roles of Gibberellins in Plant Growth and Development

Promotion of Seed Germination

☆ **Break Dormancy**: Gibberellins are crucial for breaking seed dormancy and promoting germination. In many seeds, gibberellins help activate the enzymes necessary for breaking down stored starches (like amylase) in the seed, providing energy for the growing seedling.

☆ **Triggering Enzyme Production**: During germination, gibberellins trigger the production of hydrolytic enzymes (*e.g.*, α-amylase) in the aleurone layer of the seed. These enzymes break down the starch reserves in the endosperm, releasing sugars that are used to fuel the growth of the embryo.

Stem Elongation (Cell Elongation and Division)

- ☆ **Growth Promotion**: Gibberellins are known for their role in promoting stem elongation and internode length. They stimulate cell elongation and cell division in the stem, resulting in increased plant height. This is particularly important during the early stages of growth when the plant needs to reach sunlight and outcompete other plants for light resources.
- ☆ **Dwarf Mutants**: Gibberellins also control plant stature. Certain plant mutants that lack gibberellin production (called dwarf mutants) have stunted growth due to the inability to elongate stems and other tissues.

Flowering and Flower Development

- ☆ **Flower Induction**: In some plants, gibberellins promote flowering by stimulating the transition from vegetative growth to reproductive growth. For example, in long-day plants, gibberellins can help induce flowering under appropriate light conditions.
- ☆ **Sexual Expression**: Gibberellins can also influence the sex expression in some plants. In certain species, gibberellin treatment can promote the development of male flowers (*e.g.*, in cucumbers) or female flowers (*e.g.*, in papaya), affecting fruit production.

Fruit Development and Ripening

- ☆ **Fruit Set**: Gibberellins are involved in the development of seedless fruits. In seedless grape cultivars, for example, gibberellin treatment promotes the growth of fruit without the need for pollination, resulting in seedless berries.
- ☆ **Fruit Elongation**: In some fruits, gibberellins are used in agricultural practices to promote the size and elongation of fruits, as seen in fruits like grapes, apples, and strawberries.
- ☆ **Delayed Ripening**: Gibberellins have also been shown to delay fruit ripening in certain species by inhibiting the production of ethylene, a plant hormone that regulates ripening. This effect is sometimes utilized in post-harvest treatments to extend the shelf life of certain fruits.

Leaf Expansion

- ☆ Gibberellins promote the expansion of leaves by enhancing both cell elongation and division. This process contributes to the development of large leaves that can capture more light for photosynthesis, which is crucial for plant growth and energy production.

Vascular Tissue Differentiation

- ☆ **Xylem and Phloem Development**: Gibberellins can promote the differentiation of vascular tissue (xylem and phloem), which is essential for water and nutrient transport. This contributes to overall plant growth, particularly in the development of the stem and root systems.

Interaction with Other Hormones

- ☆ Gibberellins interact with other plant hormones, such as auxins, cytokinins, and abscisic acid (ABA), to regulate various growth processes. For example:
- ☆ **Auxin and Gibberellin Interaction**: In stem elongation, gibberellins work in tandem with auxins. Auxins help direct the growth of the plant, while gibberellins promote cell elongation.
- ☆ **Cytokinin and Gibberellin Interaction**: Cytokinins can promote cell division, while gibberellins induce elongation. Together, they contribute to overall tissue growth.
- ☆ **Abscisic Acid (ABA) and Gibberellin**: Gibberellins often antagonize the effects of abscisic acid, particularly in processes like seed dormancy and germination. Gibberellins help break dormancy, while ABA typically inhibits it.

Gibberellin Biosynthesis and Signaling

- ☆ **Biosynthesis**: Gibberellins are synthesized through a complex pathway in plants, starting with the precursor geranylgeranyl pyrophosphate (GGPP). The pathway involves several enzymes that produce different types of gibberellins, each with specific biological activity. The biosynthesis of gibberellins is regulated by environmental and internal signals, such as light, temperature, and nutrient availability.
- ☆ **Gibberellin Receptors and Signaling**: Gibberellins signal through a specific receptor called the GID1 receptor (Gibberellin Insensitive Dwarf 1), which is located in the cell membrane. When gibberellin binds to this receptor, it triggers a signaling cascade that leads to the degradation of DELLA proteins, which are growth inhibitors. Once the DELLA proteins are degraded, growth-promoting genes are activated, allowing the plant to grow and develop.

Agricultural Uses of Gibberellins

- ☆ **Crop Improvement**: Gibberellins are used in agriculture to improve crop yield and quality. For example, in grape production, gibberellins are applied to produce seedless varieties and larger fruits. They are also used to increase fruit size in apples and citrus fruits.

- **Germination of Difficult Seeds**: Gibberellins are often applied to seeds that have a hard seed coat or that require specific conditions to germinate, such as conifers or palm seeds.
- **Controlling Plant Height**: Gibberellins can be used to increase plant height in certain crops, which can improve light capture and photosynthesis. However, in some cases, gibberellin inhibitors are used to dwarf crops to prevent lodging (falling over) in tall plants like wheat.

Roles of Gibberellins

- **Seed Germination**: Promote the breakdown of seed reserves and activate enzymes necessary for germination.
- **Stem Elongation**: Stimulate growth in stems and increase internode length, contributing to overall plant height.
- **Flowering**: Regulate the transition from vegetative to reproductive growth and influence sex expression in some plants.
- **Fruit Development**: Promote fruit set and elongation, especially in seedless fruit varieties.
- **Leaf Expansion**: Enhance leaf size and photosynthetic capacity.
- **Vascular Tissue Differentiation**: Contribute to the formation of xylem and phloem for water and nutrient transport.
- **Interacting with Other Hormones**: Work in conjunction with hormones like auxins, cytokinins, and abscisic acid to coordinate plant growth and responses.

Abscisic Acid

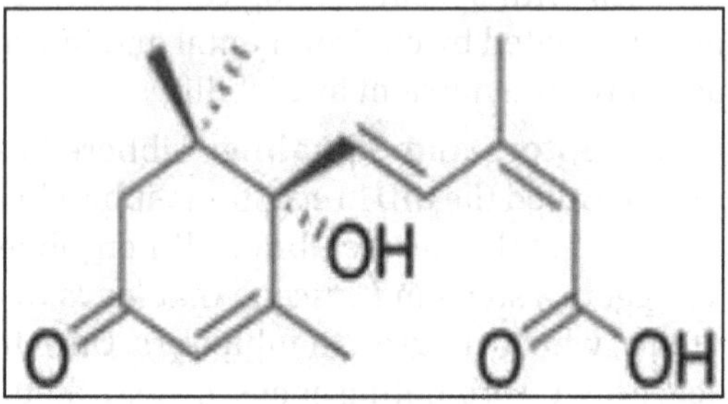

Abscisic acid (ABA) is a plant hormone that plays a crucial role in regulating various aspects of plant growth, development, and responses to environmental stress. It is often referred to as the "stress hormone" because it helps plants manage abiotic stress (*e.g.*, drought, salinity, cold) and biotic stress (*e.g.*, pathogen attack). ABA is involved in processes such as seed dormancy, stomatal closure, root development, and the regulation of growth under adverse conditions. Abscisic acid is a critical regulator of plant responses to environmental stress and plays a major role in controlling growth, development, and reproductive processes. It is especially important in managing plant responses to water stress, such as drought, and controlling processes like seed dormancy and germination. By interacting with other hormones like gibberellins and cytokinins, ABA helps plants adapt to their environment and optimize growth and survival in challenging conditions.

Roles of Abscisic Acid in Plant Growth and Development

Seed Dormancy and Germination

- **Induction of Seed Dormancy**: One of the most important roles of ABA is its involvement in the regulation of seed dormancy. High levels of ABA in seeds help maintain dormancy, preventing seeds from germinating until environmental conditions are favorable (*e.g.*, adequate moisture and temperature). This ensures that seeds do not germinate prematurely and that the plant has the best chance for survival.
- **Breaking Dormancy**: The breakdown of ABA or a reduction in its levels is required for the germination of seeds. When environmental conditions (like water availability) are right, ABA levels decrease, and other hormones like gibberellins promote the processes needed for seed germination.

Stomatal Closure and Water Conservation

- **Water Stress Response**: ABA is crucial for regulating water balance within plants, especially during times of drought or water scarcity. When a plant experiences drought, ABA is synthesized in the roots and transported to the leaves, where it triggers the closure of stomata (the small pores on the leaf surface).
- **Stomatal Regulation**: By closing the stomata, ABA reduces water loss through transpiration, which helps conserve water in the plant. This is a key mechanism that allows plants to cope with periods of water stress.

Regulation of Growth and Development

- **Inhibition of Growth**: ABA generally inhibits growth, particularly in young and growing tissues. For example, it can reduce cell elongation and inhibit shoot growth under stressful conditions. This growth inhibition helps the plant conserve energy and resources during challenging environmental conditions.
- **Regulation of Root Development**: ABA can promote root growth under drought conditions by enhancing the development of root systems that are more effective at accessing water from deeper soil layers. This is important for ensuring that the plant can continue to grow and survive during periods of water stress.

Stress Responses (Drought, Salt, and Cold Stress)

- **Drought Stress**: ABA is often referred to as the "stress hormone" because it is key in helping plants respond to drought stress. It helps plants adapt to water scarcity by inducing stomatal closure, reducing water loss, and regulating gene expression for stress tolerance.
- **Salt Stress**: ABA also plays a role in helping plants respond to salt stress, a condition that can disrupt cellular functions due to excess salts in the soil. By influencing ion transport mechanisms and gene expression, ABA helps maintain ionic balance in plant cells during salinity stress.
- **Cold Stress**: Under cold conditions, ABA helps regulate plant responses by controlling the expression of cold-responsive genes. This helps plants maintain cellular integrity and function under freezing temperatures.

Flowering and Reproductive Development

- **Regulation of Flowering**: ABA is involved in regulating the timing of flowering in response to environmental cues, such as day length and temperature. In some plants, ABA can delay flowering or regulate the transition from vegetative to reproductive growth, ensuring that plants flower under favorable conditions.
- **Reproductive Success**: ABA has also been shown to play a role in seed maturation and fruit ripening, where it influences the accumulation of storage proteins and lipids, which are essential for the embryo's survival during dormancy.

Vascular Development

- **Vascular Tissue Differentiation**: ABA is involved in the differentiation of vascular tissues (such as xylem and phloem) in the plant. This contributes to the plant's ability to transport water, nutrients, and sugars efficiently. During times of stress, ABA helps regulate the development of vascular tissues to maintain proper transport functions.

Mechanism of Action

- **ABA Signaling Pathway**: ABA signaling occurs through a well-defined pathway that involves the ABA receptors, which are located in the plasma membrane of plant cells. Once ABA binds to its receptor, a signaling cascade is initiated that leads to changes in gene expression and cellular processes.
- **DELLA Proteins and Transcription Factors**: In response to ABA, the action of DELLA proteins and transcription factors is modulated, which ultimately regulates the expression of stress-responsive genes.
- **Ion Channel Regulation**: One of the key physiological responses to ABA involves the regulation of ion channels in the stomatal guard cells. ABA

binding to its receptor activates ion pumps and channels, causing the efflux of potassium ions from the guard cells, leading to stomatal closure and reduced water loss.

ABA and Other Hormones

- **ABA vs. Gibberellins**: ABA and gibberellins (GAs) often have opposing effects in plants. While gibberellins promote growth and seed germination, ABA inhibits growth and induces dormancy. During seed germination, gibberellins help activate enzymes that break down seed reserves, while ABA prevents premature germination under unfavorable conditions.
- **ABA and Cytokinins**: Cytokinins and ABA have opposing effects on certain processes. For example, cytokinin promotes cell division and shoot growth, while ABA inhibits growth, especially under stress. The balance between these hormones regulates many aspects of plant growth and stress response.

Agricultural Applications of Abscisic Acid

- **Drought Tolerance**: ABA-based treatments are being investigated to improve drought tolerance in crops. By manipulating ABA signaling, it may be possible to enhance the plant's ability to cope with water scarcity, especially in regions facing water stress or unpredictable rainfall.
- **Seed Dormancy and Germination Control**: ABA is used in agriculture to manage seed dormancy and control germination timing. For instance, it can be applied to certain crops to prevent premature germination under unfavorable conditions, ensuring the seed will only germinate when conditions are ideal.
- **Crop Yield Enhancement**: By controlling the timing of flowering and maturation through ABA application, agriculturalists can optimize harvest times and improve crop yield.

Functions of Abscisic Acid

- **Seed Dormancy**: ABA induces seed dormancy and prevents premature germination. It must decrease for seed germination to occur.
- **Water Conservation**: ABA promotes stomatal closure during drought stress, reducing water loss through transpiration.
- **Stress Responses**: ABA helps plants respond to abiotic stresses such as drought, salt, and cold stress by modulating gene expression and maintaining cellular integrity.
- **Growth Regulation**: ABA inhibits growth under stress conditions, helping plants conserve energy and resources during unfavorable conditions.

☆ **Root and Vascular Development**: ABA can influence root growth and the formation of vascular **tissues**, aiding the plant's ability to take up water and nutrients.

☆ **Flowering and Reproduction**: ABA regulates the timing of flowering and the maturation of seeds and fruits.

Ethylene

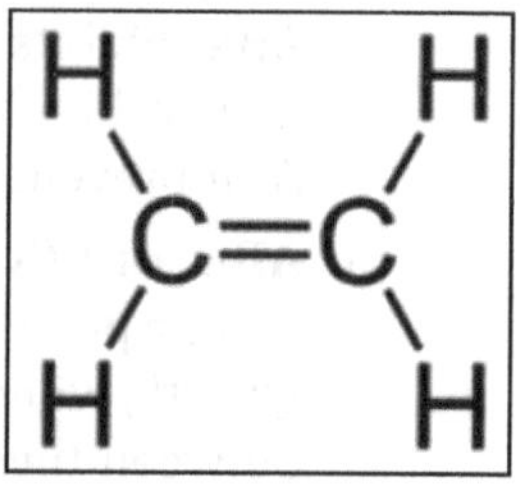

Ethylene is a unique and vital plant hormone that primarily regulates growth, development, and responses to environmental stress. It is a simple gaseous hormone that acts in very low concentrations and has a wide range of effects on plants, including promoting fruit ripening, leaf senescence (aging), and stress responses, as well as regulating other aspects of plant growth such as flowering and seed germination. Ethylene is a multifaceted and versatile plant hormone that regulates essential physiological processes, including fruit ripening, leaf senescence, stress responses, and flowering. Its role in these processes makes it indispensable for plant survival, development, and reproduction. In agriculture, ethylene is widely used to control fruit ripening and postharvest processes, making it an important tool for food production and distribution. Ethylene is produced by all plant tissues and plays an essential role in many physiological processes.

Roles of Ethylene in Plant Growth and Development

Fruit Ripening

☆ **Promotion of Ripening**: Ethylene is best known for its role in fruit ripening. It regulates the biochemical processes that convert fruits from an unripe, starch-storing state to a ripe, sugar-rich state. This includes the breakdown of cell wall components, the degradation of chlorophyll, and the synthesis of pigments (such as anthocyanins) that give fruits their color.

☆ **Ethylene as a Ripening Trigger**: In many fruits, ethylene production increases dramatically as the fruit matures. This creates a positive feedback loop, where ethylene stimulates more ethylene production, further accelerating ripening. Examples of fruits that produce ethylene include bananas, tomatoes, apples, and avocados.

☆ **Commercial Use**: In agriculture, ethylene is used to induce ripening in fruits like bananas, tomatoes, and mangoes after they have been harvested. This allows for better control of the timing of fruit ripening and post-harvest management.

Leaf Senescence and Abscission

☆ **Leaf Aging (Senescence)**: Ethylene plays a key role in leaf senescence–

the aging process of leaves. As leaves age, ethylene production increases, leading to the breakdown of cellular structures, chlorophyll degradation, and the eventual leaf drop (abscission).

- ☆ **Abscission (Leaf, Flower, and Fruit Drop)**: Ethylene is involved in the regulation of abscission zones in plants, which are areas where plant tissues like leaves, flowers, and fruits are separated from the plant. Ethylene promotes the production of enzymes that break down cell wall components in these areas, allowing for the separation of the tissue. This process is important for the plant's overall health, as it helps in shedding old or damaged parts, which can conserve resources and protect the plant.

Stress Responses and Adaptation

- ☆ **Response to Abiotic Stress**: Ethylene is produced in response to environmental stress conditions such as drought, flooding, mechanical damage, and extreme temperatures. When a plant is exposed to stress, ethylene helps trigger a stress response that prepares the plant to cope with the adverse conditions. For instance, ethylene-induced leaf senescence can allow a plant to conserve resources during periods of stress.
- ☆ **Wounding and Injury**: Ethylene is produced in response to wounding or mechanical damage (*e.g.*, by herbivores or physical damage). This promotes the healing process, as well as defense responses such as the synthesis of defense-related proteins and wound-induced growth in certain parts of the plant.
- ☆ **Flooding**: Ethylene is particularly important in waterlogged conditions, where it helps regulate the plant's aerenchyma (air-filled tissue that facilitates gas exchange). Ethylene promotes the development of this tissue, which helps the plant survive in flooded environments by improving oxygen availability.

Flowering and Germination

- ☆ **Flowering**: In certain plant species, ethylene plays a role in regulating flowering. For example, ethylene promotes flower opening in some species, while in others, it is involved in the transition from vegetative to reproductive growth. In some plants, ethylene is required to induce flowering, while in others, it may delay the process.
- ☆ **Seed Germination**: Ethylene has been implicated in seed germination, particularly in seeds that are sensitive to environmental stresses. Ethylene helps break seed dormancy in certain species, promoting the initiation of germination when conditions are favorable.

Root Growth and Development

- ☆ **Adventitious Root Formation**: Ethylene plays a role in the formation of adventitious roots–roots that develop from non-root tissues, such as stems or leaves. This is important in processes like vegetative propagation (*e.g.*, in cuttings) and responses to stress (*e.g.*, drought).
- ☆ **Root Elongation**: While ethylene can inhibit root elongation in certain conditions, it can also stimulate root growth under some circumstances, especially in response to mechanical injury or stress. Ethylene promotes the growth of lateral roots and root hairs, helping the plant to increase nutrient and water uptake.

Cell Elongation and Growth Regulation

- ☆ **Inhibition of Cell Elongation**: Ethylene can have an inhibitory effect on cell elongation in certain plant tissues, particularly in response to stress. This effect is seen in processes like stem thickening and root swelling, which are adaptive responses to physical damage or mechanical stress.
- ☆ **Hypocotyl Elongation**: In some plants, ethylene plays a role in regulating hypocotyl elongation during seedling growth, particularly under shade or crowding conditions. Ethylene helps seedlings grow towards light or expand in environments where space is limited.

Interaction with Other Plant Hormones

- ☆ **Ethylene and Auxin**: Ethylene interacts with auxins in regulating plant growth and development. For example, in apical dominance, auxins inhibit the growth of lateral buds while ethylene helps break dormancy and allows lateral buds to grow. In fruit ripening, ethylene and auxin levels coordinate to regulate the timing and extent of ripening.
- ☆ **Ethylene and Cytokinins**: Ethylene works in opposition to cytokinins in some processes. For example, while cytokinins promote cell division and growth, ethylene may inhibit cell division and promote senescence and abscission.
- ☆ **Ethylene and Gibberellins**: Ethylene and gibberellins (GAs) can have complementary or opposing effects. For example, while gibberellins promote stem elongation and growth, ethylene can inhibit it under stress conditions.

Ethylene Biosynthesis and Signaling Pathway

- ☆ **Biosynthesis**: Ethylene is produced from the amino acid methionine through a pathway that involves the intermediate S-adenosylmethionine (SAM). The key enzyme in this pathway is 1-aminocyclopropane-1-carboxylate (ACC) synthase, which converts SAM into ACC, the immediate precursor of ethylene.

- **Signaling Pathway**: Ethylene signals through a receptor-mediated pathway. Ethylene binds to ethylene receptors in the endoplasmic reticulum, which in turn activates the ETHYLENE INSENSITIVE 2 (EIN2) protein and other downstream signaling components. This leads to the activation of ETHYLENE RESPONSE FACTORS (ERFs) that regulate the expression of ethylene-responsive genes.

Agricultural and Commercial Uses of Ethylene

- **Fruit Ripening Control**: Ethylene is widely used in the commercial ripening of fruits. For instance, in banana ripening chambers, controlled exposure to ethylene gas accelerates the ripening process to synchronize fruit ripening for marketing purposes.
- **Postharvest Management**: Ethylene inhibitors, like 1-methylcyclopropene (1-MCP), are used to slow down the ripening and senescence of fruits and flowers during storage, thus extending shelf life and reducing waste.
- **Flowering Control in Crops**: Ethylene is used to induce or control flowering in certain crops, such as pineapples and chrysanthemums, which require specific ethylene concentrations to trigger flowering.

Gravitropism, Thigmotropism and other Tropisms

Tropisms are directional growth responses of plants to environmental stimuli. These responses allow plants to adapt to their surroundings, optimizing their ability to capture resources like light, water, and nutrients, or to avoid harm. Tropisms are controlled by growth regulators (hormones) such as auxins and involve differential growth rates in different parts of a plant. Tropisms can be classified based on the type of stimulus that triggers the response. Among the most well-known tropisms are gravitropism, thigmotropism, and several others, each responding to different environmental cues.

Gravitropism (Geotropism)

Gravitropism is the growth response of plants to gravity. It enables plants to orient their growth in response to the pull of gravity, ensuring that their roots grow downward into the soil (positive gravitropism) and their shoots grow upward toward the light (negative gravitropism).

- **Positive Gravitropism**: The growth of roots downward into the soil in response to gravity. Roots are sensitive to gravity and typically grow in the direction of the gravitational pull. This ensures that the plant anchors itself properly and can access water and nutrients from the soil.
- **Negative Gravitropism**: The growth of shoots, stems, and leaves in the opposite direction (upward), away from the pull of gravity. This allows the plant to grow toward light and ensures its leaves and flowers are positioned for optimal photosynthesis and pollination.

- **Mechanism**: The statocytes, specialized cells in the root cap and shoot, contain amyloplasts (starch-storing organelles) that settle under the influence of gravity. This settling provides the plant with a directional signal, which is translated into a growth response. Auxins play a key role in gravitropism by promoting elongation in cells on the opposite side of the gravitational pull, leading to differential growth.

Thigmotropism

Thigmotropism is the growth response of plants to mechanical touch or physical contact with objects. It allows plants to adapt to mechanical stimuli such as wind, touch, or the physical presence of another object.

- **Positive Thigmotropism**: In some plants, thigmotropism results in growth toward or around the object that is being touched. A classic example of positive thigmotropism is the tendrils of climbing plants, like peas or vines, which wrap around supports such as trellises, fences, or other plants to help the plant climb.
- **Negative Thigmotropism**: In other cases, plants may grow away from the object that they contact. This is observed in some roots that avoid obstacles in the soil.
- **Mechanism**: When a plant comes into contact with a solid object, it triggers mechanical signaling pathways that result in differential growth. In tendrils, for example, the cells on the side of the tendril opposite the point of contact grow faster, causing the tendril to curl around the object. Auxins and gibberellins are involved in this process, mediating the growth response to mechanical stimuli.

Phototropism

Phototropism is the growth response of plants to light. Plants are able to detect the direction of light and adjust their growth to maximize exposure to light, which is essential for photosynthesis.

- **Positive Phototropism**: The growth of plant shoots toward light. This ensures that the plant maximizes its exposure to sunlight for photosynthesis. This is most commonly observed in stems, where the plant grows toward the light source.
- **Negative Phototropism**: Some roots exhibit negative phototropism, growing away from light and into the dark soil, where they can anchor the plant and absorb water and nutrients.
- **Mechanism**: Phototropism is mediated by the hormone auxin. When light is perceived by the plant, auxin moves to the darker side of the plant. This uneven distribution of auxin promotes cell elongation on the shaded side, causing the plant to bend toward the light source. This phenomenon is

regulated by photoreceptors that detect light, such as phytochromes and cryptochromes.

Hydrotropism

Hydrotropism is the growth response of plant roots to water gradients. Roots can detect the presence of water in the soil and grow toward areas of higher moisture concentration.

- **Positive Hydrotropism**: The growth of roots toward areas of higher water availability. This is particularly important during periods of drought or when the plant needs to access water in deeper soil layers.
- **Mechanism**: The exact mechanisms behind hydrotropism are still being studied, but it is believed that roots sense changes in the water potential or osmotic pressure in the soil. Growth hormones, including auxins and gibberellins, may play a role in directing root growth toward water sources.

Chemotropism

Chemotropism is the growth response of plants to chemicals. This can involve either the attraction or repulsion of plant growth toward or away from specific chemicals in the environment.

- **Positive Chemotropism**: This can be observed in processes like pollination, where pollen tubes grow toward the ovule in response to chemical signals released by the pistil. Similarly, some plant roots may grow toward beneficial minerals or nutrients like nitrogen or phosphorus.
- **Negative Chemotropism**: In some cases, plants may grow away from harmful substances such as toxins or pollutants. For example, plant roots may avoid areas with high concentrations of toxic heavy metals or salts.
- **Mechanism**: The plant detects the chemical gradient in the surrounding environment, and signaling pathways are activated that result in growth toward or away from the chemical stimulus. This process often involves complex receptor-ligand interactions and hormone signaling.

Thigmomorphogenesis

Thigmomorphogenesis is a more general, non-directional growth response of plants to mechanical stimuli such as wind, touch, or other physical forces. This type of tropism refers to how plants alter their overall growth patterns in response to touch or mechanical stress.

- **Wind or Mechanical Stress**: When plants experience frequent mechanical stress (like wind or contact with objects), they often develop shorter, thicker stems and stronger root systems. This response helps the plant withstand physical forces by reinforcing its structural integrity.

☆ **Mechanism**: The response involves changes in gene expression that trigger the production of cell-wall strengthening proteins, lignin, and other structural compounds. Additionally, mechanical stress can also increase the production of ethylene, which plays a role in adjusting plant growth patterns under stress.

Nastic Movements

While not a tropism (since it is not directionally dependent on the stimulus), nastic movements are another form of plant response to stimuli. These are non-directional, rapid movements that occur in response to specific environmental cues, such as temperature, light, or touch.

Examples of Nastic Movements:

☆ **Thigmonasty**: Rapid movement in response to touch, such as the closing of mimosa leaves when touched.

☆ **Nyctinasty**: Movements in response to light and darkness, such as the closing of flowers or the folding of leaves at night in some species (*e.g.*, sensitive plant).

☆ **Mechanism**: Nastic movements are typically controlled by changes in turgor pressure within specific cells (like pulvini), rather than directional growth.

Seed Germination and Dormancy

Seed Germination and Dormancy are two crucial stages in the life cycle of a plant, deeply influencing its growth and survival. Seed germination is the process by which a seed develops into a new plant, typically triggered by the right environmental conditions, such as water, temperature, and oxygen. Germination begins when a seed absorbs water, causing it to swell and soften. This triggers enzymatic activity, leading to the breakdown of stored food reserves, which provides energy for the growing embryo. The seed coat ruptures, allowing the radicle (embryonic root) to emerge and anchor the plant into the soil, followed by the shoot or plumule that pushes upward toward the surface. Essential to this process is the presence of proper light and temperature conditions, which vary by plant species. Some seeds require a period of stratification (cold treatment) or scarification (physical damage to the seed coat) to overcome dormancy before they can germinate.

Seed dormancy, on the other hand, is a mechanism that prevents seeds from germinating under unfavorable environmental conditions. Dormancy ensures that seeds do not sprout prematurely, which could lead to the failure of the seedling due to lack of resources or harsh conditions. This state of rest is regulated by various factors, including hormonal signals, primarily abscisic acid (ABA), which inhibits germination. Dormancy can be broken by specific environmental cues such as temperature changes, light exposure, or after the seed coat is mechanically damaged, allowing the seed to absorb water and begin the germination process. Dormancy types vary; physiological dormancy is where metabolic processes are temporarily suspended, and morphological dormancy occurs when the seed embryo is underdeveloped at the time of dispersal. The breaking of dormancy is often a response to favorable environmental conditions, signaling the seed that it is an appropriate time to begin the growth cycle, leading to the successful establishment of a new plant.

Aging and Senescence in Plants

Aging and senescence are natural processes in plants that involve the gradual decline of cellular function and the eventual death of plant tissues. These processes are critical for the plant's life cycle, as they influence growth, reproduction, and resource allocation. Aging in plants refers to the overall accumulation of molecular damage and changes in cellular function over time. It is a gradual process that affects various parts of the plant, including roots, stems, leaves, and flowers. Aging is influenced by both intrinsic genetic factors and extrinsic environmental factors, such as temperature, light, and nutrient availability. As plants age, they may exhibit reduced photosynthetic efficiency, slowed growth, and a decline in the ability to respond to stress, which impacts their overall vitality and productivity. The accumulation of damage to cellular components, such as DNA, proteins, and lipids, over time contributes to the aging process. Additionally, plants experience oxidative stress, which results from the accumulation of reactive oxygen species (ROS), further accelerating the aging process.

Senescence, often seen as the final stage of aging, is a regulated process where the plant's cells undergo systematic breakdown, resulting in the dismantling of cellular structures. Senescence in plants is a form of programmed cell death, often initiated in response to environmental signals like drought, extreme temperatures, or nutrient depletion. In leaves, senescence is a highly coordinated process that involves the degradation of chlorophyll, proteins, and other cellular components, making nutrients like nitrogen and phosphorus available for recycling and transport to other parts of the plant, particularly the seeds. Senescence is controlled by various plant hormones, including ethylene, abscisic acid (ABA), cytokinins, and gibberellins. Ethylene, for example, plays a key role in regulating leaf senescence and fruit ripening, while ABA is associated with stress-induced senescence. In many plants, senescence occurs naturally in older leaves, flowers, or fruits after they have fulfilled their functions in photosynthesis, reproduction, or nutrient storage. However, senescence is also a mechanism that helps plants conserve resources during times of stress or unfavorable growing conditions. In some cases, controlled senescence can be beneficial to the plant, allowing it to recycle nutrients and redirect energy to reproductive structures. Ultimately, aging and senescence are adaptive processes that ensure a plant's survival, reproduction, and efficient resource management.

Chapter 6

Transport in Plants

Introduction

Transport in plants is a vital process that ensures the movement of water, minerals, and nutrients from the soil to various parts of the plant, as well as the distribution of the products of photosynthesis (mainly sugars) from the leaves to other tissues. The main structures responsible for transport are the xylem and phloem, two types of vascular tissues. Xylem primarily transports water and dissolved minerals upward from the roots to the stems and leaves, using the process of transpiration, where water is drawn up through the plant due to the evaporation of water from the leaf surfaces. This process also creates a pull that helps move water upward against gravity. Phloem, on the other hand, is responsible for transporting the products of photosynthesis, particularly sugars, from the leaves (where they are produced) to other parts of the plant for growth, storage, and energy use. This transport occurs through a process known as translocation, which relies on pressure flow mechanisms that move the sugary solution from areas of high concentration (the leaves) to areas of lower concentration (such as the roots and developing fruits). The efficiency of these transport systems is crucial for the overall health and growth of the plant, allowing it to take in essential resources from the environment and distribute them effectively to support various metabolic functions. Moreover, these transport mechanisms are influenced by environmental factors such as light, temperature, and soil water availability, which can impact a plant's ability to absorb and move water and nutrients.

Principles of Transport in Plants

The principles of transport in plants are governed by several mechanisms that ensure the efficient movement of water, nutrients, and photosynthetic products throughout the plant. These principles are based on physical processes such as diffusion, osmosis, active transport, and pressure-flow mechanisms. Below are the key principles that drive transport in plants:

- **Diffusion:** Diffusion is the movement of molecules from an area of higher concentration to an area of lower concentration. In plants, this principle is crucial for the movement of gases like oxygen and carbon dioxide during respiration and photosynthesis. Oxygen moves from the cells where it is used (in roots or leaves) to the surrounding tissues where its concentration is lower, and carbon dioxide moves from the outside air into the leaves where it is needed for photosynthesis.
- **Osmosis:** Osmosis is the movement of water molecules from a region of lower solute concentration (such as pure water) to a region of higher solute concentration through a semi-permeable membrane. In plants, osmosis plays a key role in water uptake by the roots from the soil. The root cells have a higher concentration of solutes than the surrounding soil water, so water enters the root cells by osmosis. This movement of water is essential for maintaining cell turgidity, which helps plants stay rigid and upright.
- **Active Transport:** Active transport is the movement of ions or molecules across a membrane against their concentration gradient, requiring energy (usually from ATP). In plants, active transport is used to absorb essential minerals like potassium, nitrate, and phosphate from the soil into the roots, where they are present in lower concentrations. These ions are pumped into root cells using protein pumps in the cell membranes, ensuring the plant has access to necessary nutrients.
- **Capillarity:** Capillarity is the ability of water to move upward through narrow tubes or capillaries due to the adhesive forces between water molecules and the walls of the xylem vessels and the cohesive forces between the water molecules themselves. This principle helps water to rise from the roots to the leaves, though it alone is not enough to move water over long distances in plants. Capillarity works in conjunction with other mechanisms like transpiration and cohesion.
- **Transpiration:** Transpiration is the process of water evaporation from the aerial parts of the plant, mainly through small pores called stomata on the leaves. As water evaporates from the leaf surface, it creates a negative pressure that pulls more water upward from the roots through the xylem vessels. This is known as the "transpiration stream" and is one of the primary drivers of water transport in plants. Transpiration also helps in cooling the plant and maintaining turgor pressure within cells.
- **Cohesion and Adhesion:** Cohesion refers to the attraction between water molecules, which allows them to stick together and form a continuous column of water in the xylem. Adhesion refers to the attraction between

water molecules and the walls of the xylem vessels. Together, cohesion and adhesion contribute to the capillary action and the upward movement of water through the plant.

- **Pressure Flow (Phloem Transport):** The transport of sugars and other organic compounds in plants occurs through the phloem tissue. The pressure-flow hypothesis explains how this process works. When sugars are actively loaded into the phloem from the photosynthetic tissues (usually the leaves), they create a high osmotic pressure in the sieve tubes of the phloem. This high pressure forces the sugary solution to flow from the source (where sugars are produced) to the sink (where sugars are consumed or stored). This flow is facilitated by differences in pressure between the source and sink regions.
- **Root Pressure:** Root pressure is a phenomenon where water moves into the roots and creates pressure within the xylem vessels. This pressure can force water and minerals upward, especially in smaller plants. While root pressure plays a role in water movement, it is not as significant in large plants or trees, where transpiration and cohesion dominate the transport process.
- **Bulk Flow:** Bulk flow refers to the mass movement of water and dissolved substances in response to pressure gradients. In the xylem, bulk flow moves water from the roots to the leaves under the influence of the negative pressure created by transpiration. Similarly, in the phloem, bulk flow moves sugars from areas of high pressure (sources) to areas of low pressure (sinks).
- **Water Potential:** Water potential is a measure of the potential energy of water in a system. It determines the direction of water movement in plants. Water moves from regions of higher water potential to regions of lower water potential. Water potential is influenced by both solute concentration and pressure. In the plant, the water potential gradient helps determine the movement of water through the plant tissues, from the soil to the roots, through the xylem, and into the leaves.

These principles work together to ensure that plants can efficiently absorb water and nutrients from the soil, transport them throughout the plant, and move the products of photosynthesis to areas where they are needed for growth, storage, and reproduction. Understanding these processes is essential for comprehending how plants survive and grow in different environments.

Water Transport: The Xylem

The **xylem** is a type of vascular tissue responsible for the transport of water and dissolved minerals from the roots to the rest of the plant, primarily the leaves. This process is essential for various plant functions, including photosynthesis, nutrient transport, and maintaining cell turgidity. The xylem plays a central role in maintaining the plant's internal hydration, enabling it to grow, produce energy, and remain structurally stable. Water transport through the xylem is a highly coordinated and complex process that involves several physical and biochemical mechanisms.

Structure of the Xylem

The xylem tissue is composed of several specialized cell types that work together to facilitate the movement of water. These include:

- ☆ **Tracheids**: Tracheids are elongated, tapering cells with thick walls that contain lignin, a substance that makes them rigid and waterproof. They have small openings called pits, which allow water to pass from one cell to the next. Tracheids are present in all vascular plants, including ferns and gymnosperms (like conifers), where they perform water transport.
- ☆ **Vessel Elements**: Vessel elements are shorter and wider than tracheids and are more specialized for water transport. They stack on top of each other to form long tubes, known as vessels, with perforated end walls, allowing for efficient movement of water through the plant. Vessel elements are the primary water-conducting cells in angiosperms (flowering plants), and their larger diameter and specialized structure enable faster water flow compared to tracheids.
- ☆ **Xylem Fibers**: Xylem fibers are long, narrow, and lignified cells that provide structural support to the xylem tissue. These fibers help the plant maintain rigidity and prevent collapse under internal pressure.
- ☆ **Xylem Parenchyma**: These are living cells involved in the storage of water, minerals, and nutrients. They can also participate in the lateral movement of water through the plant and help in the repair of damaged xylem tissue.

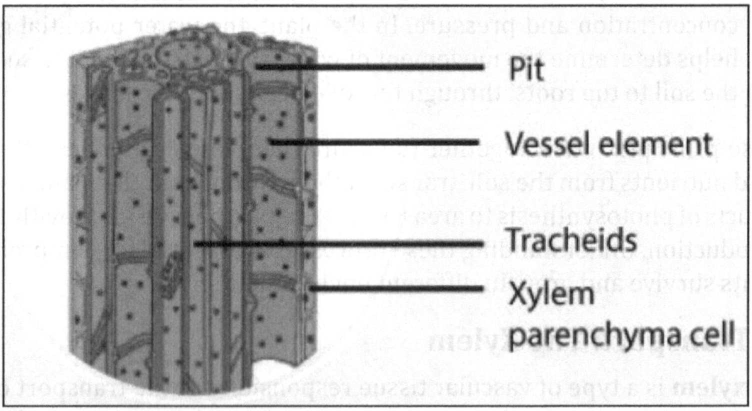

Water Transport Mechanisms in the Xylem

Water transport in plants through the xylem is a highly efficient and vital process that ensures the plant remains hydrated and able to perform critical functions such as photosynthesis and nutrient transport. The xylem's structure, combined with physical processes like transpiration, cohesion, and root pressure, allows water to move from the roots to the leaves, supporting the plant's overall growth and health. Understanding the mechanisms of xylem transport is key to studying plant physiology and the ways plants adapt to their environment.

The movement of water through the xylem is driven by a combination of physical processes, including transpiration, cohesion, adhesion, and root pressure.

Transpiration-Cohesion-Tension Mechanism

This is the primary mechanism driving water transport through the xylem. It involves three key processes:

- ☆ **Transpiration**: The evaporation of water from the plant's leaves, primarily through small pores called stomata, creates a loss of water from the leaf surface. As water evaporates from the leaf, it creates a negative pressure (tension) in the leaf's xylem vessels.
- ☆ **Cohesion**: Water molecules exhibit a property called cohesion, where they stick to each other due to hydrogen bonds. This cohesion helps maintain a continuous column of water within the xylem, allowing the tension generated by transpiration to pull water upward from the roots through the plant.
- ☆ **Adhesion**: Water also adheres to the walls of the xylem vessels due to the hydrogen bonds between water molecules and the cellulose in the xylem cell walls. This adhesive force helps counteract gravity and supports the upward movement of water.

As water evaporates from the stomata (a process called evapotranspiration), it pulls water upward from the roots to the leaves, creating a continuous flow of water through the plant. This is the cohesion-tension theory, and it accounts for the movement of water through the xylem in most plants.

Root Pressure

Root pressure is another mechanism that can assist in water movement, though it is more significant in smaller plants and plays a less important role in tall trees. Root pressure occurs when water enters the roots through osmosis, causing a build-up of pressure in the xylem vessels. This pressure can push water upward through the plant, especially in the early morning when transpiration is low.

Root pressure is most noticeable in certain conditions, such as high soil moisture or during the night when transpiration is reduced. However, in larger plants and trees, transpiration-driven cohesion and tension are the primary mechanisms responsible for water movement.

Water Movement Through the Xylem

Water moves through the xylem in a continuous stream from the roots to the leaves via a series of interconnected vessels. The process involves several steps:

- **Uptake of Water by the Roots**: Water enters the roots through osmosis from the surrounding soil. The root cells have a higher concentration of solutes than the soil solution, so water moves into the roots to balance the concentration.
- **Movement through the Root Cortex**: Once inside the root, water moves through the root cortex, which is the layer of cells between the epidermis and the vascular cylinder. Water can move along two pathways:
- The **apoplastic pathway**, where water moves through the cell walls and intercellular spaces.
- The **symplastic pathway**, where water moves through the cytoplasm of cells via plasmodesmata (tiny channels between cells).
- **Entry into the Xylem**: Once the water reaches the vascular cylinder in the root, it enters the xylem vessels. This process is facilitated by the endodermis, a layer of cells that regulates the flow of water and minerals into the vascular tissue.
- **Movement Up the Plant**: Once inside the xylem, water moves upward due to the negative pressure created by transpiration, the cohesion between water molecules, and the adhesive properties of water to the xylem walls. The xylem vessels form a continuous pipeline, allowing water to travel from the roots through the stem and into the leaves.
- **Water Evaporation from the Leaves**: Water exits the plant through the stomata in the leaves in the form of water vapor, completing the transpiration process. The loss of water from the leaves creates the necessary tension to pull more water up through the plant from the roots.

Factors Affecting Water Transport in the Xylem

Several factors can influence the efficiency of water transport through the xylem:

- **Temperature**: High temperatures increase the rate of transpiration, which in turn enhances the upward movement of water. However, excessive heat can cause excessive water loss and dehydration of the plant.
- **Humidity**: Low humidity in the air increases transpiration, while high humidity reduces the rate of water loss from the plant.
- **Soil Water Availability**: Adequate water in the soil is essential for efficient water uptake by the roots. Water stress or drought can severely impair water transport in the xylem.

- ☆ **Water Quality**: The presence of minerals and solutes in water can affect its movement through the plant. High concentrations of certain ions can block the xylem vessels and reduce water flow.
- ☆ **Xylem Vessel Condition**: The integrity of xylem vessels is crucial for the uninterrupted movement of water. Blockages, air bubbles (embolism), or damage to the xylem can disrupt water flow.

Phloem Transport: Translocation of Sugar

Phloem transport is the process by which organic compounds, primarily sugars (mainly sucrose), are moved from source tissues (where they are produced or stored) to sink tissues (where they are utilized, stored, or converted). This movement of sugars and other organic molecules is essential for plant growth, development, and energy distribution. Unlike the unidirectional movement of water through the xylem, phloem transport (also called translocation) occurs in both directions, depending on the plant's needs. The phloem is the vascular tissue responsible for the transport of these substances throughout the plant. Phloem transport, or translocation, is an essential process in plants, responsible for moving the products of photosynthesis (especially sugars) from the source to the sink. This process is complex and involves the active loading and unloading of sugars, the creation of pressure gradients, and the bulk flow of water and solutes. The efficient movement of sugars allows plants to grow, develop, and respond to environmental changes. Understanding the mechanisms of phloem transport is crucial for agricultural practices and in studying plant physiology, particularly in terms of crop productivity, nutrient allocation, and overall plant health.

Structure of the Phloem

The phloem tissue is composed of several specialized cell types that work together to ensure efficient transport of sugars and other organic molecules. These include:

- ☆ **Sieve Elements (Sieve Tubes)**: Sieve elements are the main cells involved in the transport of sugars. They are long, cylindrical cells that are connected end-to-end, forming continuous tubes known as sieve tubes. The ends of sieve elements have sieve plates, which are perforated structures that allow the flow of water and dissolved substances from one cell to the next. In angiosperms (flowering plants), sieve tube elements are living cells, but they lack a nucleus and many other organelles, which is compensated for by the presence of companion cells.
- ☆ **Companion Cells**: Companion cells are closely associated with sieve tube elements and are responsible for managing the metabolic activities of the sieve tube elements. These cells have a full set of organelles, including a nucleus, which they use to support the sieve tube elements. Companion cells facilitate the loading and unloading of sugars into the sieve tubes,

and they help maintain the pressure gradient necessary for phloem transport.

- **Phloem Parenchyma**: These are living cells that help store and transport nutrients, including sugars and other organic compounds, within the phloem. They play a role in the lateral movement of substances between sieve tubes and other plant tissues.
- **Phloem Fibers**: Phloem fibers are elongated cells that provide structural support to the phloem tissue. They help the plant maintain its rigidity and prevent collapse under pressure.

The Process of Translocation: How Sugars Move through the Phloem

Translocation is an active and energy-dependent process that relies on both pressure gradients and active transport mechanisms. The movement of sugars through the phloem follows a source-to-sink pathway, with the source being the location of sugar production (typically leaves or storage organs) and the sink being areas of sugar utilization or storage (such as roots, stems, flowers, or fruits).

Loading of Sugars into Phloem (Source Region)

The first step in translocation is the loading of sugars (primarily sucrose) into the sieve tubes from source tissues (usually the leaves). This process is active and requires energy.

- **Active Transport of Sucrose**: In the leaves, sucrose is produced through photosynthesis and must be loaded into the sieve tube elements for transport. This occurs in the mesophyll cells of the leaf, where sucrose is actively transported into companion cells and then into the sieve tube elements. The energy for this process comes from ATP generated by cellular respiration. Sucrose is loaded into the sieve tubes through specialized transporter proteins in the cell membranes.
- **Osmosis and Pressure Build-up**: As sucrose accumulates in the sieve tube elements, it increases the solute concentration in the phloem. This lowers the water potential (osmotic potential) in the sieve tubes, causing water from the adjacent xylem to move into the phloem by osmosis. The influx of water creates positive pressure within the phloem, which is critical for driving the flow of sap (the solution of water and dissolved nutrients) through the sieve tubes. This pressure is sometimes referred to as turgor pressure.

Movement of Sugars Through the Phloem (Transport Region)

Once sucrose is loaded into the sieve tubes, it moves through the plant towards the sink regions, driven by a pressure gradient.

- **Pressure Flow Hypothesis**: The movement of sugars through the phloem is explained by the pressure flow hypothesis. This theory posits that water moves from areas of high pressure (the source, where sugars are loaded) to areas of low pressure (the sink, where sugars are unloaded), carrying the dissolved sugars with it.

- ❖ In the source region (*e.g.*, leaves), the active loading of sugars into the sieve tubes creates high turgor pressure in the phloem.
- ❖ As a result, the solution of water and sugars moves down the pressure gradient towards areas of lower pressure, such as roots or developing fruits, where the sugars are consumed or stored.

This bulk flow of water and solutes within the phloem is driven primarily by the osmotic pressure differences and the turgor pressure created by the loading of sucrose in the source and the unloading of sucrose at the sink.

Unloading of Sugars at the Sink (Sink Region)

In the sink tissues (*e.g.*, roots, tubers, flowers, or developing fruits), the sucrose is unloaded from the phloem to be used or stored.

- ✫ **Active Transport and Diffusion**: The unloading process often requires active transport. In sink cells, sucrose is actively removed from the sieve tubes and transferred into the surrounding cells. This can be accomplished by sucrose transporters located in the cell membranes of sink tissues. In some cases, sucrose is converted to starch (*e.g.*, in roots and tubers) or used for energy production in cells via cellular respiration.
- ✫ **Water Loss and Pressure Drop**: As sucrose is unloaded at the sink, the solute concentration in the phloem decreases, causing water to move out of the sieve tubes and into the surrounding cells. This reduces the turgor pressure in the phloem at the sink, creating a pressure gradient that allows continued flow of the sugar solution from the source.

Recirculation of Water through the Xylem

After water leaves the phloem at the sink, it often returns to the xylem, where it is recirculated to the source regions. This ensures that the plant maintains the necessary water balance and continues to support the transport of water, nutrients, and sugars.

Factors Affecting Phloem Transport

The efficiency and direction of phloem transport are influenced by several factors:

- ✫ **Light and Photosynthesis**: The production of sugars in the source regions (mainly the leaves) is directly influenced by photosynthesis. A reduction in light (*e.g.*, during the night or under shading) can decrease the amount of sugar produced, affecting the rate of translocation.
- ✫ **Temperature**: Higher temperatures can increase the rate of translocation by enhancing the activity of enzymes involved in sugar production and

transport. However, extreme temperatures can damage plant cells and reduce the efficiency of phloem transport.

- ☆ **Water Availability**: Adequate water availability is crucial for maintaining the osmotic gradient that drives phloem transport. Drought conditions can lead to reduced water flow, impairing the pressure gradient and slowing down the movement of sugars.
- ☆ **Hormonal Regulation**: Plant hormones such as auxins, cytokinins, and gibberellins can influence the rate of sugar transport and distribution, often by regulating the growth and development of sink tissues.
- ☆ **Phloem Blockages or Damage**: Physical damage to the phloem, such as embolism (air bubbles in the sieve tubes) or mechanical injury, can disrupt translocation. Additionally, any factors that inhibit the function of companion cells can also impair sugar loading and unloading.

Membrane Transport Mechanisms: Active and Passive Transport

In cells, the movement of substances across the plasma membrane is essential for maintaining homeostasis and supporting cellular functions. This movement can occur through two primary types of mechanisms: passive transport and active transport. These processes are fundamental to the life of a cell, allowing it to take in nutrients, expel waste products, and maintain appropriate internal conditions. Membrane transport mechanisms, including active and passive transport, are fundamental to cellular function and homeostasis. Passive transport occurs without the expenditure of energy and relies on concentration gradients to move molecules across the membrane, while active transport requires energy to move substances against their gradients. Both processes are essential for the proper functioning of cells, tissues, and organs, enabling them to take in nutrients, expel waste, and maintain internal conditions necessary for life. Understanding these mechanisms is crucial in areas ranging from cell biology to medicine, as disruptions in transport can lead to various diseases and disorders.

Passive Transport

Passive transport is the movement of molecules across the plasma membrane without the use of energy (ATP). The driving force for passive transport is the concentration gradient – substances move from areas of higher concentration to areas of lower concentration until equilibrium is reached. There are several forms of passive transport:

- ☆ **Simple Diffusion:** Simple diffusion is the movement of small or nonpolar molecules (such as oxygen, carbon dioxide, or lipids) across the plasma membrane. These molecules can pass directly through the phospholipid bilayer, following the concentration gradient. Example: Oxygen entering cells from the surrounding environment and carbon dioxide leaving cells as a waste product.

☆ **Facilitated Diffusion:** Facilitated diffusion involves the movement of larger or polar molecules (like glucose or ions) through the membrane via protein channels or carriers. These proteins are embedded in the membrane and help the molecules cross without expending energy. **Example**: Glucose entering cells through a glucose transporter protein or ions moving through ion channels.

❖ **Osmosis:** Osmosis is a specific type of facilitated diffusion that refers to the movement of water molecules through a selectively permeable membrane. Water molecules move from areas of lower solute concentration to areas of higher solute concentration. **Example**: Water entering plant root cells or red blood cells absorbing water in the body.

Active Transport

Active transport is the movement of substances against their concentration gradient, from areas of lower concentration to areas of higher concentration. This process requires the expenditure of energy in the form of ATP to move molecules across the membrane. Active transport is crucial for maintaining cellular functions, such as nutrient uptake and ion balance.

☆ **Primary Active:** TransportIn primary active transport, energy from ATP hydrolysis is directly used to pump molecules against their concentration gradient via pump proteins. These proteins typically work by changing shape after ATP binding, allowing them to transport ions across the membrane. Example: The sodium-potassium pump (Na^+/K^+ ATPase), which pumps sodium ions (Na^+) out of the cell and potassium ions (K^+) in, maintaining crucial gradients for nerve function and cell homeostasis.

☆ **Secondary Active Transport (Cotransport):** Secondary active transport does not directly use ATP; instead, it uses the energy from ion gradients created by primary active transport. In secondary active transport, ions (such as Na^+ or H^+) move down their concentration gradient, and this gradient drives the transport of other molecules (like glucose or amino acids) against their concentration gradient. Example: The sodium-glucose cotransporter (SGLT) in the intestines or kidneys, which uses the Na^+ gradient established by the sodium-potassium pump to bring glucose into cells.

☆ **Bulk Transport (Vesicular Transport):** Bulk transport involves the movement of large particles, liquids, or even whole cells through the plasma membrane. This process requires energy and typically involves the formation of **vesicles**, which transport materials into (endocytosis) or out of (exocytosis) the cell. **Example**:

❖ **Endocytosis**: The uptake of large molecules (*e.g.*, cholesterol or nutrients) via vesicles (*e.g.*, receptor-mediated endocytosis).

- **Exocytosis**: The expulsion of waste materials or secretion of substances like hormones from the cell.

Table 6.1. Comparison of Active and Passive Transport

Feature	*Passive Transport*	*Active Transport*
Energy Requirement	No energy required (ATP)	Requires energy (ATP)
Movement Direction	Down the concentration gradient (high to low)	Against the concentration gradient (low to high)
Transport Proteins	Often requires channel or carrier proteins	Requires pump proteins and sometimes secondary transport proteins
Examples	Simple diffusion, facilitated diffusion, osmosis	Sodium-potassium pump, glucose transporters (secondary transport), vesicular transport
Types of Molecules	Small, nonpolar molecules (gases), water, ions	Ions, large molecules, nutrients, and waste materials

Membrane Transport Mechanisms

- **Simple Diffusion:** An image showing small molecules like oxygen moving through the lipid bilayer from a region of high concentration to low concentration.
- **Facilitated Diffusion:** An image showing a protein channel allowing ions or larger molecules, like glucose, to pass through the membrane.
- **Osmosis:** A diagram showing water molecules moving through a semipermeable membrane from a region of lower solute concentration to a region of higher solute concentration.
- **Sodium-Potassium Pump (Primary Active Transport):** A diagram of the sodium-potassium pump actively transporting three sodium ions out of the cell and two potassium ions into the cell using ATP.
- **Secondary Active Transport:** An image showing the sodium-glucose cotransporter utilizing the Na^+ gradient (created by the sodium-potassium pump) to transport glucose into the cell.
- **Vesicular Transport:** A diagram of endocytosis showing a vesicle engulfing large particles or molecules from the extracellular fluid, or exocytosis expelling materials like neurotransmitters or waste from the cell.

Symplast and Apoplast Pathways

In plants, water and solutes must move from the roots, through the stems, and into the leaves or other tissues. This movement occurs via two main pathways in the plant's vascular system: the symplast pathway and the apoplast pathway. These pathways refer to the routes through which water and dissolved substances travel as they move from one cell to another, either through the cell membrane or between cells. Both pathways are essential for nutrient and water transport, but they differ in their mechanisms and the way they interact with plant cell structures.

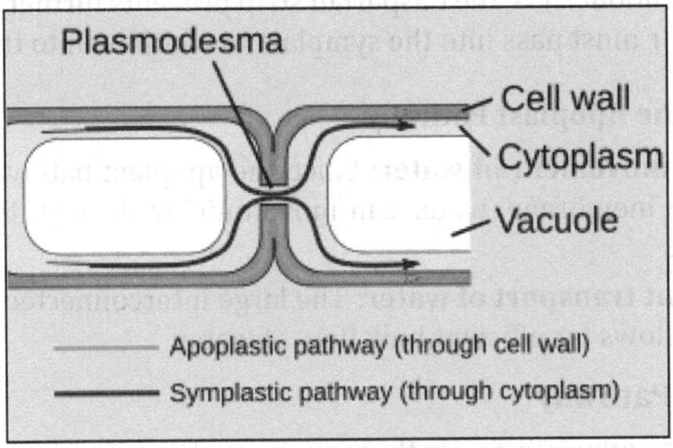

Figure 6.1

The Apoplast Pathway

The apoplast pathway involves the movement of water and solutes outside of the plant's cells, through the cell walls and intercellular spaces. The apoplast consists of the interconnected network of cell walls, intercellular spaces, and the xylem (the vascular tissue that carries water and dissolved minerals upward from the roots). Water moves freely through this extracellular matrix, which does not involve passing through the plasma membrane of the plant cells.

Features of the Apoplast Pathway

- ✰ **Movement through cell walls and intercellular spaces**: Water and solutes move between the cells via the continuous network of cellulose fibers that make up the cell walls. These walls are permeable to water, allowing it to flow through easily.
- ✰ **No membrane crossing**: Water does not need to cross the plasma membrane in the apoplast pathway, which allows for the rapid movement of water and dissolved substances.
- ✰ **Non-selective transport**: Water in the apoplast pathway may carry dissolved solutes (such as minerals and ions) without much regulation or control. However, once water reaches the endodermis (the innermost layer of cells in the root cortex), the movement is controlled, as the endodermal cells contain the Casparian strip (a barrier that forces water to move through the symplast for selective uptake).

Steps in the Apoplast Pathway

- ☆ Water enters the plant through the roots from the soil.
- ☆ Water moves through the cell walls and intercellular spaces of the root cortex and into the vascular tissue.
- ☆ At the endodermis, the Casparian strip prevents further apoplastic flow, so water must pass into the symplast to continue into the xylem.

Advantages of the Apoplast Pathway

- ☆ **Rapid movement of water**: Since the apoplast pathway bypasses the plasma membrane, water can move quickly through the extracellular spaces.
- ☆ **Efficient transport of water**: The large interconnected network of cell walls allows for efficient bulk flow of water.

The Symplast Pathway

The symplast pathway refers to the movement of water and solutes through the cytoplasm of plant cells, interconnected by plasmodesmata–microscopic channels that connect the cytoplasm of adjacent plant cells. In the symplast pathway, water moves through the cell membrane, into the cytoplasm, and from one cell to the next via plasmodesmata, allowing for a more selective and regulated form of transport.

Features of the Symplast Pathway

- ☆ **Movement through the cytoplasm**: Water and dissolved solutes travel through the cytoplasm of the plant cells. Plasmodesmata allow substances to move directly from one cell to another, bypassing the cell walls.
- ☆ **Selective transport**: Unlike the apoplast, the symplast pathway allows for more selective movement of molecules. Cells can regulate the uptake and movement of ions and nutrients through active transport mechanisms, offering greater control over what enters the plant.
- ☆ **Crossing the plasma membrane**: Water must first cross the plasma membrane of a cell before it can enter the symplast pathway. This is usually achieved through aquaporins (specialized water channels) that facilitate the movement of water into the cytoplasm.

Steps in the Symplast Pathway

- ☆ Water enters the plant cells at the root epidermis via osmosis.
- ☆ The water moves through the cytoplasm of the root cells, crossing cell membranes and passing through plasmodesmata from one cell to another.
- ☆ Water continues its journey through the root cortex, moving toward the vascular tissue.
- ☆ In the vascular tissue (xylem), water enters the vessels for long-distance transport to the stems and leaves.

Advantages of the Symplast Pathway

- ☆ **Regulation and selectivity**: The symplast pathway allows plants to control the uptake of ions, nutrients, and water, providing a level of precision in what is allowed to pass through the plant.
- ☆ **Cell-to-cell communication**: The plasmodesmata that connect cells also enable the movement of small signaling molecules, which can regulate processes like growth and response to stress.

Table 6.2. Comparison of the Apoplast and Symplast Pathways

Feature	*Apoplast Pathway*	*Symplast Pathway*
Route of Movement	Through cell walls and intercellular spaces	Through the cytoplasm of cells via plasmodesmata
Membrane Crossing	Does not involve crossing the plasma membrane	Requires crossing the plasma membrane to enter the cytoplasm
Selectivity	Non-selective; water and solutes move freely	Selective; movement can be regulated by membrane proteins
Speed of Transport	Faster due to bulk flow in extracellular spaces	Slower, as water moves through cytoplasm and plasmodesmata
Control of Water and Solutes	Limited control; solutes may flow freely until the endodermis	Greater control; solutes can be actively transported
Barrier	Casparian strip at the endodermis restricts further apoplastic flow	Plasma membranes in each cell act as a barrier until water enters the symplast

Interactions Between Apoplast and Symplast Pathways

In a plant, both pathways are interconnected and often work together to facilitate the movement of water and nutrients. For example:

- ☆ In the roots, water initially moves through the apoplast pathway through the cell walls and intercellular spaces. However, when the water reaches the endodermis, the Casparian strip blocks the apoplastic flow. At this point, water is forced to move into the symplast pathway to continue its journey into the vascular system (the xylem).
- ☆ The symplast pathway allows for selective uptake of minerals and ions, while the apoplast pathway allows for rapid bulk flow of water. Together, they ensure that the plant can efficiently transport water, nutrients, and other solutes across various tissues while maintaining control over what enters the vascular system.

The apoplast and symplast pathways are two essential routes for water and solute movement in plants, each with its distinct characteristics and advantages. The apoplast pathway is fast and non-selective, moving substances through the extracellular spaces, while the symplast pathway is more controlled and selective, moving substances through the cytoplasm and plasmodesmata. The combined use of these pathways allows plants to efficiently manage the uptake of water, minerals, and nutrients while maintaining cellular control and homeostasis. Both pathways are crucial for the plant's overall water and nutrient transport system, ensuring proper growth and function.

The Role of Aquaporins in Water Movement

Aquaporins are specialized membrane proteins that facilitate the rapid and selective transport of water molecules across cell membranes. These channels are crucial for the regulation of water movement in plant cells, playing a key role in various physiological processes, including water absorption in roots, transpiration in leaves, and maintaining water homeostasis throughout the plant. Aquaporins are found in all living organisms, including plants, animals, and microorganisms, but their specific function and regulation can vary depending on the organism and tissue type. In plants, aquaporins are especially important because they allow for the fine-tuned movement of water, ensuring that the plant maintains appropriate hydration levels under changing environmental conditions. They are involved in processes such as nutrient transport, turgor pressure regulation, and the movement of water across tissues in response to environmental signals. Aquaporins are crucial for the movement of water in plants, playing an essential role in water uptake, transport, regulation of turgor pressure, and transpiration. By forming highly selective channels in the plasma membrane, aquaporins allow water to move rapidly across cell membranes, enabling plants to maintain hydration and regulate water use efficiently. The activity of aquaporins is tightly controlled by environmental signals, hormones, and post-translational modifications, allowing plants to adapt to varying water conditions and stresses. Understanding aquaporin function is important for improving plant water use efficiency, particularly in the face of changing climate conditions and water scarcity.

Structure of Aquaporins

Aquaporins belong to a family of proteins known as major intrinsic proteins (MIPs). The structure of an aquaporin consists of:

- ☆ **Amino acid sequence**: Aquaporins are composed of hydrophobic amino acid sequences that span the membrane, creating a pathway for water molecules to pass through.
- ☆ **Water channel**: The protein forms a central hydrophilic channel that allows water molecules to move through the lipid bilayer of the cell membrane.

- **Selective permeability**: Aquaporins are highly selective, allowing water molecules to pass through the membrane, but typically preventing the passage of ions or other small solutes (though some aquaporins may allow other molecules like glycerol or urea to pass in specific circumstances).

Aquaporins typically form tetramers, meaning that four aquaporin protein subunits can come together to form a functional water channel.

Function of Aquaporins in Water Movement

Aquaporins facilitate water transport through the plant cell membrane by acting as highly selective channels. Their primary function is to increase the permeability of the cell membrane to water, allowing water to move in and out of cells at a much faster rate than would occur through simple diffusion across the lipid bilayer.

- **Water Absorption in Roots:** Aquaporins play an essential role in water uptake by plant roots. When water from the soil enters the root, it must cross several cell layers before reaching the vascular tissue (xylem) for transport to other parts of the plant. Aquaporins in the root cell membranes enable water to move across the cell walls and through the plasma membranes more efficiently.
  - In root hair cells, aquaporins allow for rapid water uptake from the soil.
  - In the root cortex, aquaporins facilitate the movement of water toward the vascular tissue.
  - In endodermal cells, aquaporins assist in regulating the flow of water into the xylem, where it is transported upward to the stems and leaves.
- **Water Transport through Plant Tissues:** Once water enters the plant, aquaporins are crucial in the movement of water through plant tissues. In vascular tissues such as the xylem, water is transported through the plant by bulk flow, but aquaporins are still involved in regulating the movement of water across cell membranes within the tissues.
  - Aquaporins in xylem parenchyma cells control the movement of water from xylem vessels into surrounding cells and across vascular bundles.
  - In leaf mesophyll cells, aquaporins help facilitate water movement from the xylem to the cells responsible for photosynthesis, and also play a role in regulating transpiration, the process by which water vapor is lost from the plant to the atmosphere.
- **Transpiration and Regulation of Water Loss:** Aquaporins are involved in regulating transpiration–the evaporation of water from the leaf surface, primarily through stomata. While the majority of water loss occurs through the stomata, aquaporins in guard cells (cells surrounding

stomata) play a crucial role in controlling the water content in these cells. By regulating the movement of water into or out of the guard cells, aquaporins help control stomatal opening and closing, which in turn affects transpiration and the plant's water use efficiency.

- When water enters the guard cells through aquaporins, the cells swell, causing the stomatal pore to open. Conversely, when water exits, the cells shrink, causing the stomata to close.
- This regulation is important for minimizing water loss during dry conditions while allowing for gas exchange necessary for photosynthesis.

☆ **Turgor Pressure Regulation:** Aquaporins also play a role in maintaining turgor pressure, the pressure exerted by the cell's contents on the cell wall. Turgor pressure is essential for maintaining cell structure and function, especially in non-woody tissues.

- Aquaporins in vacuolar membranes (tonoplasts) help maintain water movement into and out of the vacuole, which plays a key role in regulating turgor pressure.
- Turgor pressure also helps drive the growth of plant cells by enabling the cells to expand as water is absorbed into the vacuole.

Regulation of Aquaporin Activity

Aquaporin activity is highly regulated in response to various environmental and physiological signals. The plant can adjust the expression and activity of aquaporins to optimize water uptake and distribution under different conditions.

Environmental Signals

☆ **Water Availability**: When water availability is low (*e.g.*, during drought stress), aquaporins can be downregulated to prevent excessive water loss. In contrast, when water availability is high, aquaporins can be upregulated to facilitate water uptake.

☆ **Soil Moisture and Osmotic Conditions**: Aquaporins respond to changes in soil water potential and osmotic conditions, helping the plant adjust its water uptake accordingly.

Hormonal Regulation

☆ **Abscisic Acid (ABA)**: This hormone is involved in regulating plant responses to water stress. ABA often leads to the closure of stomata and can also influence the expression of aquaporins, reducing water loss during drought conditions.

- **Cytokinins and Gibberellins**: These hormones, which regulate growth, can also influence aquaporin expression, particularly in tissues like the roots, where water uptake is critical for cell expansion and growth.

Post-translational Modifications

Aquaporins can be modulated by post-translational modifications such as phosphorylation, which can affect their activity, localization, and interaction with other proteins. Phosphorylation typically affects the opening and closing of the water channels.

Aquaporins and Plant Water Stress

Aquaporins are essential in helping plants cope with water stress, which can occur due to factors such as drought, soil salinity, or high temperatures. Under stress conditions, plants may alter the number or activity of aquaporins to reduce water loss, maintain turgor pressure, or adjust the uptake of water from the soil.

- In drought conditions, the plant may reduce the activity of aquaporins to conserve water.
- In saline environments, aquaporins may play a role in managing the uptake of water while minimizing the absorption of toxic salts.
- Cold stress may also affect aquaporin activity, as ice formation can disrupt water transport, and aquaporins may help the plant adapt to freezing conditions by facilitating intracellular water movement.

Chapter 7

Plant Signaling and Communications

Introduction to Plant Signaling

Plant signaling is a complex and dynamic process by which plants detect, process, and respond to a variety of environmental, developmental, and internal cues. Unlike animals, plants do not have a nervous system to coordinate responses, but they have evolved sophisticated signaling mechanisms to ensure survival and growth in a constantly changing environment. These signaling pathways allow plants to sense their surroundings, such as light, temperature, gravity, water, nutrients, and the presence of pathogens, and then trigger appropriate physiological responses. The study of plant signaling is a multidisciplinary field that intersects plant biology, molecular biology, biochemistry, and genetics, aiming to understand how plants communicate internally and externally to adapt to their environment. At the core of plant signaling are various signaling molecules and receptors that initiate and mediate cellular responses. These molecules include hormones, such as auxins, cytokinins, gibberellins, abscisic acid, and ethylene, which are often referred to as plant growth regulators. These hormones act as messengers, traveling through plant tissues to regulate processes like growth, development, fruit ripening, and stress responses. In addition to hormones, plants use secondary messengers like calcium ions, cyclic nucleotides, and reactive oxygen species (ROS) to amplify and transduce signals within cells.

One of the key features of plant signaling is its ability to integrate various signals to produce a coordinated response. For instance, in response to environmental stress such as drought or pathogen attack, plants can activate a network of genes that help to conserve water, enhance defense mechanisms, or repair damaged tissues. The ability to "perceive" and respond to these signals often involves complex interactions between receptor proteins on the plant cell surface and intracellular signaling pathways. These pathways can activate transcription factors that initiate changes in gene expression, ultimately influencing plant behavior. Moreover, plant signaling is not limited to internal processes but also involves communication with other organisms. For example, plants can release volatile organic compounds to signal neighboring plants about the presence of herbivores, triggering a defensive response in the surrounding vegetation. Similarly, plants engage in mutualistic relationships with microorganisms in the soil, exchanging signals that regulate processes like nutrient uptake or disease resistance. This interplant signaling and interaction with microbes underline the interconnectedness of plant systems within ecosystems. The study of plant signaling has far-reaching implications for agriculture, forestry, and environmental science. Understanding how plants respond to stressors, diseases, and changing climates can help in developing crops that are more resilient to these challenges. Advances in plant signaling research could also lead to innovations in crop management, pest control, and sustainable farming practices. By unraveling the intricate molecular pathways of plant communication, scientists are unlocking new possibilities for improving food security and enhancing the health of our planet's ecosystems.

Signal Transduction Pathways in Plants

Signal transduction in plants refers to the process by which external or internal signals are received by the plant and translated into a cellular response. This process is fundamental for plants to adapt to their environment, regulate growth, and respond to various stimuli, such as light, temperature, gravity, mechanical stress, and biotic factors like pathogens or herbivores. Signal transduction involves a sequence of molecular events, starting with the recognition of a signal by specific receptors, followed by the activation of intracellular signaling molecules, and culminating in a physiological response that alters gene expression, enzyme activity, or cellular metabolism. In plants, these pathways are often highly conserved across species and can integrate multiple signals to orchestrate appropriate responses. Signal transduction pathways are essential for plant survival, enabling them to sense and respond to a broad array of environmental and internal signals. From receptor activation to changes in gene expression, these pathways orchestrate a variety of plant processes, including growth, development, stress responses, and defense mechanisms.

A deeper understanding of these pathways not only sheds light on the fundamental biology of plants but also holds promise for improving agricultural practices, enhancing crop resilience, and developing strategies for sustainable farming in the face of climate change and other global challenges.

- ☆ **Receptors and Signal Perception:** The first step in signal transduction is the perception of an external or internal signal, often by specialized proteins called receptors. These receptors are typically located in the cell membrane, but can also be found in the cytoplasm or the nucleus. The types of signals they detect include mechanical pressure, light, hormones, or chemical molecules produced by pathogens. Upon recognizing a signal, these receptors undergo a conformational change, which initiates the transduction process.

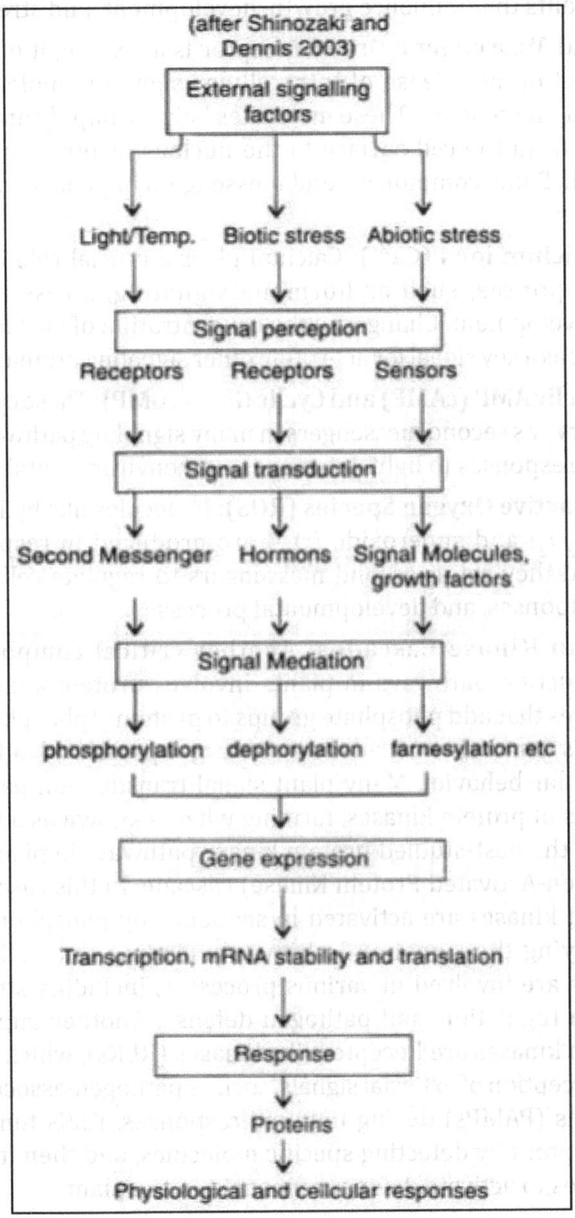

Figure 7.1. Signal Transduction Pathways in Plants.

For example:

- **Photoreceptors** like phytochromes and cryptochromes detect light and help regulate processes like seed germination, photomorphogenesis, and flowering.
- **Hormone receptors** bind plant growth regulators (like auxins, gibberellins, or abscisic acid), triggering a cascade of downstream events that influence growth, development, and stress responses.

☆ **Second Messengers:** Once a receptor is activated, it often triggers the production or release of intracellular signaling molecules known as second messengers. These molecules help to amplify and propagate the signal from the cell surface to the nucleus or other organelles within the cell. Some common second messengers in plant signal transduction include:

- **Calcium ions (Ca^{2+})**: Calcium plays a crucial role in various plant responses, such as hormone signaling, stress responses, and development. Changes in the concentration of Ca^{2+} in the cytoplasm act as a key signal for activating other signaling proteins and pathways.
- **Cyclic AMP (cAMP) and Cyclic GMP (cGMP)**: These cyclic nucleotides serve as second messengers in many signaling pathways, particularly in responses to light, hormones, and environmental stresses.
- **Reactive Oxygen Species (ROS)**: Molecules like hydrogen peroxide (H_2O_2) and superoxide (O_2^-) are produced in response to stress, and they act as second messengers to regulate cell death, defense responses, and developmental processes.

☆ **Protein Kinase Cascades:** Another critical component of signal transduction pathways in plants involves protein kinases, which are enzymes that add phosphate groups to proteins (phosphorylation). This process can activate or deactivate target proteins, leading to changes in cellular behavior. Many plant signal transduction pathways involve a series of protein kinases, forming what is known as a kinase cascade. One of the best-studied protein kinase pathways in plants is the MAPK (Mitogen-Activated Protein Kinase) cascade. In this cascade, a series of protein kinases are activated in sequence by phosphorylation events, amplifying the signal and ultimately leading to a cellular response. MAPKs are involved in various processes, including stress responses, growth regulation, and pathogen defense. Another important class of protein kinases are Receptor-like kinases (RLKs), which are involved in the perception of external signals, such as pathogen-associated molecular patterns (PAMPs) during immune responses. RLKs function similarly to receptors by detecting specific molecules, and then initiating kinase cascades to activate defense responses in the plant.

- **Transcriptional Regulation and Gene Expression:** As a result of these signaling events, plants often alter their gene expression to adjust to the environment. Signal transduction pathways activate transcription factors, which are proteins that bind to DNA and regulate the expression of specific genes. This regulation can lead to changes in cell metabolism, growth, and development. For instance, in response to stress (like drought or pathogen attack), the plant activates specific transcription factors, such as ABF (ABA-responsive element-binding factors), which trigger the expression of genes involved in stress tolerance mechanisms, like the production of osmoprotectants or defensive proteins. Similarly, in response to light, transcription factors like HY5 (elongated hypocotyl help control the expression of genes involved in photomorphogenesis.
- **Hormone Signaling Pathways:** Plant hormones are key players in signal transduction pathways, acting as primary messengers that help plants respond to both internal and external stimuli. Each hormone works through a unique signaling pathway, and several hormones may act in concert to regulate growth and responses to environmental conditions.
  - **Auxin signaling**: Auxins are involved in processes like cell elongation, root development, and phototropism. The auxin signal is transmitted by receptors in the cell membrane, such as TIR1, which interacts with the auxin molecule to trigger a cascade of events leading to changes in gene expression.
  - **Gibberellin signaling**: Gibberellins regulate processes like seed germination, stem elongation, and flowering. The gibberellin receptor activates a signaling cascade that involves the degradation of repressor proteins, which allows the transcription of growth-promoting genes.
  - **Abscisic acid (ABA) signaling**: ABA is crucial in response to abiotic stresses like drought. ABA binds to its receptor, leading to the activation of protein kinases and transcription factors that regulate stress tolerance genes.
  - **Ethylene signaling**: Ethylene plays a role in fruit ripening, senescence, and response to mechanical stress. Ethylene receptors in the endoplasmic reticulum mediate changes in gene expression through a signaling pathway that involves the CTR1 kinase, which inactivates the EIN2 protein, leading to the activation of downstream transcription factors.
- **Integration of Signals and Crosstalk**: One of the most fascinating aspects of plant signal transduction is the integration of multiple signaling pathways. Plants often experience several concurrent environmental stresses (*e.g.*, drought, heat, or pathogen attack), and their ability to coordinate a response is critical for survival. Signaling pathways are not

isolated but interact in a process known as crosstalk, where the output of one pathway can influence the activation or repression of another. For example, the plant hormone jasmonic acid, involved in defense responses, can crosstalk with salicylic acid signaling, which plays a role in pathogen resistance. Similarly, light signaling pathways can interact with hormone pathways to modulate growth and development in response to environmental light conditions.

- **Feedback and Homeostasis:** Finally, many plant signal transduction pathways involve feedback mechanisms that regulate the intensity and duration of the signal. Negative feedback loops can help return the system to a baseline state after the response is triggered, preventing overreaction. For instance, the ABA signaling pathway is tightly controlled to prevent excessive water loss under stress conditions, ensuring that the plant doesn't dehydrate too much.

Plant Hormones in Signal Transduction

Plant hormones, also known as phytohormones, play a crucial role in regulating various physiological processes in plants, including growth, development, and responses to environmental stimuli. Signal transduction refers to the process by which a cell translates an external signal (such as the presence of a hormone) into a functional response, typically involving a series of molecular events. Plant hormone signal transduction pathways are sophisticated and highly integrated, with each hormone influencing a variety of physiological processes through intricate signaling networks. These pathways involve receptor-ligand interactions, second messengers, protein kinases, transcription factors, and feedback regulation, enabling plants to adapt to internal and external cues for optimal growth and survival.

Major Plant Hormones and Their Involvement in Signal Transduction Pathways

Auxins (IAA - Indole-3-acetic acid)

- **Function**: Auxins regulate cell elongation, root development, and differentiation, and are involved in tropisms (*e.g.*, phototropism and gravitropism).
- **Signal Transduction**:
  - Auxins are perceived by receptors known as TIR1/AFB (Transport Inhibitor Response/AUXIN SIGNALING F-Box).
  - Binding of auxin to TIR1/AFB triggers the degradation of Aux/IAA proteins, which are repressors of auxin-responsive genes.
  - This leads to the activation of ARF (Auxin Response Factors), which bind to promoter regions of target genes to regulate their expression.
  - This pathway regulates processes such as elongation, root formation, and differentiation.

Cytokinins

- **Function**: Cytokinins promote cell division and differentiation, delay leaf senescence, and are involved in shoot formation.
- **Signal Transduction**:
  - Cytokinins are recognized by receptors located in the plasma membrane, including histidine kinase receptors.
  - Cytokinin binding activates a signaling cascade through a two-component signaling system that involves phosphorylation of response regulators (RRs).
  - This leads to the activation or repression of cytokinin-responsive genes, controlling processes like cell division and differentiation.

Gibberellins (GA)

- **Function**: Gibberellins are involved in promoting seed germination, stem elongation, flower and fruit development, and overcoming dormancy.
- **Signal Transduction**:
  - Gibberellins bind to the GID1 receptor (Gibberellin Insensitive Dwarf 1).
  - This binding causes a conformational change in GID1, enabling it to interact with DELLA proteins, which normally repress GA responses.
  - GA-GID1-DELLA interactions lead to the degradation of DELLA proteins via the proteasome pathway, thus removing the repression and allowing the expression of genes involved in growth and development.

Abscisic Acid (ABA)

- **Function**: ABA is essential for stress responses, including drought and salinity tolerance, and it plays a key role in seed dormancy and germination.
- **Signal Transduction**:
  - ABA is perceived by the PYR/PYL family of receptors, which are part of the START domain proteins.
  - Upon ABA binding, the PYR/PYL receptors inhibit PP2C (Protein Phosphatase 2C), which normally represses the SnRK2 family of protein kinases.
  - The activation of SnRK2s leads to the phosphorylation of downstream targets, including transcription factors that regulate stress-related genes.
  - ABA signaling also involves the regulation of stomatal closure and gene expression changes related to stress responses.

Ethylene

- ☆ **Function**: Ethylene regulates fruit ripening, flower senescence, leaf abscission, and response to environmental stress.
- ☆ **Signal Transduction**:
  - ❖ Ethylene is perceived by the ETHYLENE RECEPTOR family located in the endoplasmic reticulum.
  - ❖ In the absence of ethylene, these receptors inhibit the CTR1 kinase, preventing the activation of downstream signaling.
  - ❖ When ethylene binds to the receptor, it inactivates CTR1, leading to the activation of EIN2 (Ethylene Insensitive 2) and the downstream EIN3 transcription factor, which regulates the expression of ethylene-responsive genes.
  - ❖ Ethylene signaling involves crosstalk with other hormones like auxins and jasmonates.

Jasmonates (JA)

- ☆ **Function**: Jasmonates are involved in defense responses, particularly against herbivores and pathogens, and regulate processes like flower development and seed germination.
- ☆ **Signal Transduction**:
  - ❖ The primary receptor for jasmonates is COI1 (Coronatine-Insensitive 1), a component of the SCF complex.
  - ❖ Jasmonic acid (JA) binds to COI1, leading to the ubiquitination and degradation of JAZ (Jasmonate ZIM-domain) proteins, which are repressors of MYC2 transcription factors.
  - ❖ Degradation of JAZ proteins releases MYC2, which activates the expression of genes involved in defense responses.

Brassinosteroids (BRs)

- ☆ **Function**: Brassinosteroids regulate cell elongation, vascular differentiation, stress tolerance, and reproductive development.
- ☆ **Signal Transduction**:
  - ❖ BRs are perceived by BRI1 receptors on the plasma membrane.
  - ❖ Binding of BRs to BRI1 activates a kinase cascade that involves BAK1 (BRASSINOSTEROID-SIGNALING KINASE 1) and other kinases.
  - ❖ This cascade leads to the phosphorylation and activation of BES1/BZR1 transcription factors, which regulate the expression of BR-responsive genes.

Salicylic Acid (SA)

- **Function**: Salicylic acid is involved in plant defense, particularly in systemic acquired resistance (SAR) against pathogens.
- **Signal Transduction**:
  - SA signaling is initiated by the binding of SA to Non-expressor of Pathogenesis-Related Genes (NPR1).
  - This interaction promotes NPR1's oligomerization and translocation into the nucleus, where it regulates the expression of pathogenesis-related (PR) genes.
  - The SA-NPR1 pathway activates a defense response that enhances resistance to pathogen invasion.

Crosstalk Between Hormones

- Hormonal interactions are key to coordinating responses to environmental signals and internal cues.
  - For example, auxins and cytokinins often work together to balance cell division and differentiation in root and shoot development.
  - Ethylene and jasmonates may cooperate in defense responses, while ABA and gibberellins often antagonize each other, with ABA promoting stress responses and gibberellins promoting growth.
  - Brassinosteroids and auxins also interact to control cell growth and differentiation.

Environmental Sensing: Light, Temperature and Gravity

Plants are highly sensitive to environmental signals, and their ability to sense and respond to factors like light, temperature, and gravity is crucial for optimizing growth, reproduction, and survival. The processes through which plants detect and react to these stimuli are often highly sophisticated, involving complex signal transduction pathways, molecular receptors, and physiological responses. The environmental cues of light, temperature, and gravity play pivotal roles in regulating the growth and developmental processes of plants, from seed germination to flowering, and they ensure that plants can thrive in a changing and sometimes hostile environment. Here, we delve deeply into the mechanisms plants use to sense and respond to these key environmental factors.

Light Sensing and Photoreception

Light is arguably one of the most critical environmental factors influencing plant growth and development. Plants rely on light for photosynthesis, but they also use it as a signal to regulate various aspects of their development, such as germination, stem elongation, flowering, and even the timing of leaf senescence. Light sensing in plants is primarily mediated by specialized photoreceptors that allow plants to detect specific wavelengths of light and initiate appropriate physiological responses.

The most well-known photoreceptors in plants are phytochromes, cryptochromes, and phototropins. Each of these photoreceptors responds to different light wavelengths and has distinct roles in regulating plant growth. Phytochromes, for example, are sensitive to red and far-red light. They exist in two interconvertible forms: Pr (red light-absorbing form) and Pfr (far-red light-absorbing form). The balance between these forms allows plants to detect the light quality in their environment. Phytochromes are critical in processes like seed germination, shade avoidance, and flowering time. For example, when a plant is in shaded conditions, the ratio of red to far-red light changes, which signals the plant to elongate its stems to outgrow its neighbors and access more light.

Cryptochromes, on the other hand, are blue-light receptors that regulate processes such as circadian rhythms, leaf expansion, and flowering. These receptors help plants synchronize their biological clocks to the day-night cycle, enabling them to anticipate changes in light conditions. Phototropins, which are also blue-light receptors, mediate phototropism–the growth of plant stems and roots in response to directional light. This allows plants to grow toward light sources (a process known as positive phototropism), maximizing their ability to capture light for photosynthesis. At the cellular level, light signals are transduced through complex pathways involving secondary messengers such as calcium ions and cyclic GMP, and ultimately, these signals activate transcription factors that regulate gene expression. Through these signaling pathways, plants can adjust their growth patterns, optimize photosynthesis, and ensure that they are positioned in the most favorable light conditions.

Temperature Sensing and Thermoregulation

Temperature is another key environmental factor that significantly affects plant growth and development. Plants are highly sensitive to both temperature fluctuations and extremes, as these can have profound impacts on metabolic processes, enzyme activities, and cellular structure. The ability to sense and respond to temperature changes is vital for plants to optimize their growth and reproduction in various climates. Plants have evolved a number of mechanisms to sense temperature changes, and these mechanisms are often integrated with the plant's internal circadian clock, allowing for a coordinated response to daily or seasonal temperature cycles. One of the primary ways plants sense temperature is through a group of proteins known as thermosensors. These proteins undergo conformational changes in response to temperature fluctuations, which then trigger downstream signaling pathways. An example of this is the C-repeat binding factor (CBF) signaling pathway, which regulates cold stress responses. When plants experience cold temperatures, CBF transcription factors activate the expression of cold-responsive genes, leading to the production of proteins and metabolites that help protect the plant from freezing and other cold-related damage.

Heat shock proteins (HSPs) play a crucial role in the plant's response to high temperatures. When plants are exposed to heat stress, HSPs act as molecular chaperones, preventing the denaturation of other proteins and assisting in the repair of damaged proteins. The activation of heat shock transcription factors (HSFs) in response to high temperatures leads to the expression of HSP genes, which is a key adaptive response to thermal stress. Moreover, temperature sensing in plants is closely tied to their reproductive cycles. For example, temperature plays a significant role in determining the timing of flowering in many species. In some plants, a process called vernalization occurs, where exposure to cold temperatures during the winter induces flowering in the spring. This mechanism ensures that plants flower at the optimal time of year for successful pollination and seed production. Overall, plant temperature sensing involves a combination of thermosensors, transcription factors, and regulatory networks that allow the plant to respond to both high and low temperatures. These responses can involve adjusting metabolic pathways, altering gene expression, and activating protective mechanisms to survive temperature extremes.

Gravity Sensing and Gravitropism

Gravity is an omnipresent force in the plant environment, and the ability to sense gravity is critical for proper plant orientation and growth. This process, known as gravitropism (or geotropism), enables plants to grow roots downward into the soil (positive gravitropism) and stems upward toward the light (negative gravitropism). Gravity sensing in plants is vital for root anchorage, nutrient uptake, and optimal growth. The primary organelles responsible for gravity sensing in plants are called statocytes, which are specialized cells found in the root cap and in the shoot tissues. Inside these cells are amyloplasts (starch-filled plastids), which act as statoliths–dense particles that settle to the bottom of the cell under the influence of gravity. The movement of these statoliths within the cell is sensed by the plant as a gravitational signal. This mechanical signal is transduced into a biochemical signal that activates downstream signaling pathways, leading to changes in gene expression that guide plant growth in the correct orientation. In roots, the presence of gravity causes the statoliths to accumulate in the lower part of the root cap cells, where they activate signaling pathways that lead to differential growth. The cells on the lower side of the root elongate less than those on the upper side, causing the root to bend downward toward the gravitational pull. In stems and shoots, the process is reversed: cells on the lower side of the stem elongate more than those on the upper side, causing the stem to bend upward, away from gravity, in order to access light. The key signaling pathways involved in gravitropism include the auxin signaling pathway. Auxins are plant hormones that are crucial for growth regulation, and they play a central role in gravitropism by controlling the distribution of growth-promoting signals within the plant. When gravity is sensed, auxin distribution becomes asymmetric across the root or stem, leading to differential growth on the two sides of the organ, thereby causing the bending response. This ability to adjust growth direction in response to gravity ensures that plants can maintain proper orientation for both nutrient uptake (through roots) and photosynthesis (through leaves).

Stress Signaling: Responses to Abiotic and Biotic Stresses

Plants are constantly exposed to a variety of environmental challenges, both biotic (caused by living organisms like pathogens, herbivores, and competitors) and abiotic (caused by non-living environmental factors like extreme temperatures, drought, salinity, and heavy metals). To survive and thrive, plants have evolved sophisticated mechanisms to detect and respond to these stresses. These responses involve complex signaling networks that integrate environmental signals, activate defense mechanisms, and trigger adaptive changes in metabolism, growth, and development. The ability to respond to both abiotic and biotic stresses is critical for plant survival, growth, and reproduction, especially in the context of a changing climate and increasing biotic threats.

Abiotic Stress Responses

Abiotic stressors, including drought, extreme temperatures (heat and cold), salinity, and soil pollution, pose significant threats to plant health and productivity. These stresses can damage cellular structures, inhibit metabolic processes, and reduce growth and reproduction. To mitigate the negative effects of these environmental factors, plants rely on highly coordinated signaling pathways that involve both hormonal regulation and changes in gene expression.

- **Drought Stress:** Drought is one of the most common and severe abiotic stresses, affecting plant water availability and leading to osmotic imbalances, cellular dehydration, and oxidative stress. Plants perceive drought stress through the detection of low water potential or dehydration in tissues. This signal is primarily transduced through the abscisic acid (ABA) signaling pathway. ABA is a key plant hormone that plays a central role in drought response by regulating stomatal closure to reduce water loss and activating genes that help the plant conserve water and maintain cellular integrity. In response to drought, ABA accumulates in plant tissues and binds to receptors known as PYR/PYL proteins, which in turn activate protein phosphatases such as PP2C. This activates a cascade of signaling events involving the SnRK2 protein kinases, which trigger the expression of drought-responsive genes. These genes code for proteins that help the plant adapt to water scarcity, including dehydrins, LEA (Late Embryogenesis Abundant) proteins, and antioxidant enzymes that protect cells from oxidative damage. Furthermore, ABA signaling also induces root growth patterns that help the plant access water deeper in the soil. Reactive oxygen species (ROS) also play an essential role in drought signaling. While ROS are often considered damaging agents, they also function as secondary messengers in stress signaling. ROS trigger the activation of various transcription factors that promote stress tolerance and activate protective mechanisms, including the synthesis of osmolytes (such as proline) that stabilize cellular structures under dehydration conditions.

- ☆ **Salt Stress:** High salinity in the soil causes osmotic stress and ion toxicity, leading to reduced water uptake, cellular dehydration, and interference with cellular processes. Plants respond to salt stress by activating ion transporters, synthesizing compatible solutes, and adjusting their metabolic processes to cope with the high ion concentrations. Salt-induced signaling is mediated through the SOS (Salt Overly Sensitive) pathway, where SOS1 is a plasma membrane Na^+/H^+ antiporter that helps exclude excess sodium ions from the cytoplasm. Additionally, ABA plays a role in regulating salt stress responses. High salt concentrations activate the production of ABA, which leads to the activation of SOS2 (a protein kinase) and the subsequent phosphorylation of SOS3, which helps to modulate ion transport and promote cellular acclimation. At the molecular level, salt stress also induces the expression of genes involved in the synthesis of compatible solutes, such as trehalose and glycine betaine, which help stabilize cellular structures and maintain osmotic balance under saline conditions. In response to salt stress, plants may also adjust their root architecture, enhancing root elongation to reach water in deeper soil layers, or they may increase the production of antioxidant enzymes to counteract ROS-induced damage.
- ☆ **Temperature Stress (Heat and Cold):** Extreme temperatures, whether high or low, can have detrimental effects on plant growth and survival. Heat stress causes proteins to denature, disrupts cellular membranes, and impedes photosynthesis, while cold stress can result in membrane rigidification, freezing, and oxidative damage. In response to heat stress, plants activate heat shock proteins (HSPs), which act as molecular chaperones to protect other proteins from denaturation and assist in the refolding of damaged proteins. The primary signaling pathway in heat stress involves the activation of heat shock transcription factors (HSFs), which bind to heat shock elements in the promoters of heat shock protein genes. Additionally, plants often produce ROS in response to both heat and cold stress, which act as secondary messengers to activate stress-responsive transcription factors like DREB (Dehydration Responsive Element-Binding proteins). Under low temperatures, plants activate cold-responsive genes, including those involved in the production of cryoprotectants and lipid desaturation, to stabilize cellular membranes and protect cellular structures from freezing damage. Plants also employ osmotic adjustments and membrane restructuring to survive temperature extremes, and many species undergo a process of cold acclimation, where prior exposure to low temperatures leads to increased tolerance to freezing.

Biotic Stress Responses

Biotic stresses, caused by pathogens (bacteria, fungi, viruses), herbivores, and other plant competitors, represent another significant challenge to plant survival. These stresses trigger complex immune responses, often involving both local and systemic signaling pathways that protect plants from infection, limit the spread of pathogens, and activate defense mechanisms.

- ☆ **Pathogen Defense (PAMP-Triggered Immunity and Effector-Triggered Immunity):** Plants have evolved intricate immune systems to detect and respond to pathogen attack. The first line of defense is known as PAMP-triggered immunity (PTI), which occurs when pathogen-associated molecular patterns (PAMPs), such as bacterial flagellin or fungal chitin, are recognized by pattern recognition receptors (PRRs) on the plant cell surface. The binding of PAMPs to PRRs triggers a signaling cascade involving MAPK (mitogen-activated protein kinase) pathways, calcium ions, and ROS, which activate defense-related genes and induce a variety of responses, such as the production of antimicrobial peptides and the reinforcement of cell wall structures. However, many pathogens produce effectors–molecules that interfere with the plant's immune response. In response, plants have evolved effector-triggered immunity (ETI), where intracellular receptors (often known as NLRs, for nucleotide-binding leucine-rich repeat receptors) detect specific pathogen effectors. This leads to a more robust and localized defense response, often involving a hypersensitive response (HR), where infected cells undergo programmed cell death to limit the spread of the pathogen. Salicylic acid (SA), jasmonic acid (JA), and ethylene (ET) are key signaling molecules involved in biotic stress responses. While SA is critical for defense against biotrophic pathogens (which feed on living plant tissues), JA and ET are primarily involved in responses to necrotrophic pathogens (which kill host tissues to feed) and herbivores. Crosstalk between these hormone pathways ensures that plants can fine-tune their immune responses based on the type of biotic stress they encounter.
- ☆ **Herbivore Defense:** Herbivory by insects and other animals also triggers strong defense responses in plants. These responses include the production of toxic secondary metabolites, such as alkaloids and terpenoids, which deter herbivores from feeding, as well as the release of volatile organic compounds (VOCs) that attract natural predators of herbivores, such as parasitoid wasps. Herbivore attack also activates the jasmonic acid (JA) signaling pathway, which coordinates the production of defense proteins like proteinase inhibitors that interfere with the herbivore's digestion. Additionally, mechanical defenses such as trichomes (hair-like structures on the surface of plants) and thorns can physically deter herbivores from feeding. These physical and chemical

defense mechanisms are often tightly regulated to balance growth and defense.

Systemic Signaling: The Role of Jasmonates, Salicylates and Ethylene

Plants are constantly exposed to a wide range of environmental challenges, including abiotic and biotic stresses. To effectively cope with these stresses, plants have developed intricate systemic signaling mechanisms that allow them to coordinate their responses across different tissues. The hormonal signals involved in these systemic responses include jasmonates (JAs), salicylic acid (SA), and ethylene (ET)–three critical plant hormones that regulate defense mechanisms, growth, and development. These hormones not only mediate local responses at the site of stress but also orchestrate systemic acquired resistance (SAR), ensuring that other parts of the plant are primed to defend against further challenges. Through complex signaling pathways, plants can fine-tune their immune responses, modulate their growth and development, and communicate with other plants and organisms in their environment.

Jasmonates: Key Regulators of Defense and Stress Response

Jasmonates (JAs), particularly jasmonic acid (JA) and its bioactive derivatives, play an essential role in the plant's response to a variety of stresses, especially herbivory and necrotrophic pathogen attacks. Jasmonates, salicylic acid, and ethylene are pivotal hormones in plant systemic signaling, enabling plants to mount coordinated responses to both abiotic and biotic stresses. Jasmonates primarily regulate responses to herbivory and necrotrophic pathogens, while salicylic acid is central to defense against biotrophic pathogens and the activation of systemic acquired resistance. Ethylene acts as a versatile regulator, working with both jasmonates and salicylic acid to fine-tune the plant's immune responses, modulate growth, and orchestrate defense mechanisms. These three hormones do not function in isolation; instead, they interact in complex ways to ensure that the plant's responses are well-coordinated and appropriate for the type of stress encountered. Understanding the roles of these hormones in plant stress signaling has important implications for improving crop resistance to diseases, pests, and environmental stressors, thereby enhancing agricultural productivity and sustainability. JAs are involved in both local defense responses at the site of injury or infection and systemic signaling that primes other parts of the plant for enhanced defense.

- ☆ **Jasmonate Signaling Pathways:** Jasmonate signaling begins when JA or its derivative methyl jasmonate (MeJA) binds to the COI1 (Coronatine Insensitive 1) receptor, a component of a receptor complex involved in SCF (Skp1-Cullin-F-box) ubiquitin ligase machinery. This receptor interaction promotes the degradation of JAZ (Jasmonate ZIM-domain) proteins, which are repressors of MYC2, a key transcription factor in the JA signaling pathway. In the absence of JA, JAZ proteins suppress MYC2 activity and prevent the expression of defense-related genes. Upon JA binding, the degradation of JAZ proteins releases MYC2, allowing it to activate the transcription of genes involved in defense, including those encoding proteinase inhibitors, defensins, and other antimicrobial peptides.

- ☆ **Systemic Signaling and Priming:** The role of JAs extends beyond the immediate site of injury or pathogen attack. Once activated locally, JA signaling triggers the synthesis of systemic signals that travel through the plant to uninfected or undamaged tissues. These signals prepare distant tissues for a faster and more robust response if the stress spreads. This process is part of systemic acquired resistance (SAR), where methyl jasmonate (MeJA) is often involved as a signaling molecule that moves through the phloem to other parts of the plant, activating defense genes in a primed state, making these tissues more responsive to subsequent pathogen attacks or herbivore feeding. Moreover, JA is also involved in induced systemic resistance (ISR), a form of systemic resistance that is triggered by beneficial microorganisms (*e.g.*, certain plant growth-promoting rhizobacteria). This reinforces the idea that JAs play a central role in coordinating the plant's defense system, not only by responding to external stressors but also by preparing the entire plant for potential threats.

Salicylic Acid: Regulating Immunity and Plant Stress Resistance

Salicylic acid (SA) is a key player in plant immunity, particularly in defense against biotrophic pathogens (those that feed on living tissues) and in the regulation of systemic acquired resistance (SAR). SA is involved in the induction of local defense responses as well as systemic signaling that prepares the entire plant for future pathogen attacks.

- ☆ **SA Signaling Pathway:** SA is synthesized in response to pathogen recognition, primarily through pattern recognition receptors (PRRs) that detect pathogen-associated molecular patterns (PAMPs). The binding of these patterns by PRRs triggers pattern-triggered immunity (PTI), leading to the accumulation of SA in plant tissues. Once produced, SA activates a signaling cascade that includes the NPR1 (Non-expresser of PR genes 1) protein, a central regulator of SA-mediated defense responses. NPR1 is involved in the transcriptional activation of pathogenesis-related (PR) genes, which encode proteins with antimicrobial properties that help control pathogen growth. NPR1 also plays a key role in SAR. Following the recognition of a pathogen at a local site of infection, SA accumulates and activates NPR1, which then moves into the nucleus and interacts with various transcription factors to induce the expression of PR genes throughout the plant. This response provides protection against future pathogen infections, essentially "priming" the plant's defense system.

- **Systemic Signaling and SAR:** The systemic nature of SA signaling is particularly important in the context of SAR. Following localized infection or injury, SA is transported throughout the plant, and this movement is believed to occur via the phloem, allowing it to reach distal tissues. In these uninfected tissues, SA induces a primed state, meaning they are ready to mount a faster and more effective immune response if the pathogen spreads. This priming of defense responses enhances the plant's ability to resist further attacks and is considered a form of plant immune memory. The interplay between SA and jasmonic acid (JA) is crucial in shaping the plant's immune response. While SA is critical for defense against biotrophic pathogens, JA generally regulates defense against necrotrophic pathogens and herbivores. The balance between these two hormones determines the nature of the plant's defense response, and often, the two pathways are antagonistic. For example, SA can suppress JA-dependent defenses and vice versa, ensuring that the plant prioritizes the appropriate response based on the type of threat it faces.

Ethylene: A Versatile Hormone in Stress and Defense

Ethylene (ET) is a gaseous plant hormone that regulates a wide range of physiological processes, including growth, fruit ripening, senescence, and responses to both abiotic and biotic stresses. Ethylene plays a central role in mediating plant responses to pathogen infections, herbivore attacks, and environmental stressors, often working in coordination with JA and SA.

- **Ethylene in Biotic Stress Responses:** Ethylene is involved in the regulation of defense responses against necrotrophic pathogens, which kill plant tissues and feed on dead or dying cells. In response to such pathogens, ethylene signaling is activated and acts in concert with jasmonates to induce the expression of defense-related genes, including those encoding protease inhibitors and defensins. Ethylene also regulates the hypersensitive response (HR), a form of programmed cell death that limits the spread of infection by isolating the pathogen from healthy tissue. One key aspect of ethylene's role in plant defense is its interaction with jasmonic acid (JA). Both hormones work together in the defense response to necrotrophic pathogens and herbivores, with ethylene often modulating the levels and activity of JA. Ethylene also plays a role in activating the expression of PR genes during pathogen attacks and, like JA, may promote the production of secondary metabolites that deter herbivores or inhibit pathogen growth.
- **Crosstalk with SA and JA:** Ethylene signaling intersects with both SA and JA signaling pathways. In some cases, ethylene can synergistically enhance the actions of JA in response to necrotrophic pathogens or herbivores, while in other instances, it can antagonize SA-dependent defense mechanisms. This balance between ethylene, SA, and JA is

critical for fine-tuning the plant's defense response, ensuring that the plant mounts an effective defense that is appropriate for the nature of the stressor. In wound responses, such as those triggered by insect feeding or mechanical damage, ethylene is also involved in promoting the expression of wound-related genes, including those that produce volatile organic compounds (VOCs), which can attract natural predators of herbivores or induce defenses in neighboring plants. This underscores the role of ethylene as a key player in systemic signaling during herbivore and pathogen attacks.

Chapter 8

Photosynthesis and the Global Carbon Cycle

Introduction

Photosynthesis is a vital process that underpins the global carbon cycle, serving as a natural mechanism to regulate atmospheric carbon dioxide (CO_2) levels and thus influencing Earth's climate. In essence, photosynthesis is the process by which plants, algae, and some bacteria convert light energy, typically from the sun, into chemical energy stored in glucose, a form of sugar that organisms use for growth and energy. During this process, plants absorb CO_2 from the atmosphere and water (H_2O) from the soil, releasing oxygen (O_2) as a byproduct. This transformation is crucial because it acts as a primary means of removing CO_2 from the atmosphere, which, if left unchecked, would accumulate and contribute to the greenhouse effect, causing global warming. The glucose produced in photosynthesis serves as the foundation for energy transfer through food webs, as it provides the energy source for plants, herbivores, and ultimately carnivores. As plants grow, they also store carbon in their tissues, acting as carbon sinks that help mitigate climate change. Through the global carbon cycle, this carbon is cycled through different Earth systems – the atmosphere, biosphere, oceans, and soil – and can remain sequestered in plants, soils, or the ocean for extended periods. However, human activities, particularly deforestation and the burning of fossil fuels, have disrupted this cycle, leading to higher concentrations of atmospheric CO_2. This, in turn, exacerbates global warming and climate change. The importance of photosynthesis in the carbon cycle is underscored by its ability to counterbalance the release of CO_2 into the atmosphere. Forests, especially tropical rainforests, and oceanic phytoplankton are some of the most significant carbon sinks, playing a central role in carbon sequestration. The disruption of these ecosystems, through activities such as logging and pollution, reduces the planet's capacity to absorb CO_2, further accelerating climate change. Furthermore, climate change itself can impact the ability of plants to perform photosynthesis efficiently, as rising temperatures, droughts, and extreme weather events stress plants and reduce their carbon uptake. Addressing this challenge involves protecting and enhancing natural carbon sinks and finding ways to mitigate carbon emissions through efforts such as reforestation, sustainable agriculture, and the development of carbon capture technologies. In sum, photosynthesis is not only essential for the survival of life on Earth but also for maintaining the balance of carbon in the atmosphere and mitigating the impacts of climate change, making it a cornerstone of the global carbon cycle.

Photosynthesis and its Role in the Global Carbon Cycle

Photosynthesis is a critical process that supports life on Earth and plays an essential role in the global carbon cycle. It is the biochemical process by which green plants, algae, and certain bacteria convert light energy, typically from the sun, into chemical energy stored in glucose, a form of sugar. During this process, plants absorb carbon dioxide (CO_2) from the atmosphere and water (H_2O) from the soil, using sunlight to produce glucose ($C_6H_{12}O_6$) and release oxygen (O_2) as a byproduct. This process not only sustains plant growth and provides energy for other organisms in the food web, but also acts as a major mechanism for regulating atmospheric CO_2 levels, which is vital for maintaining Earth's climate. The global carbon cycle is the continuous movement of carbon through the atmosphere, biosphere, oceans, and geosphere. Photosynthesis is a central component of this cycle because it removes CO_2 from the atmosphere and incorporates it into living organisms. This "carbon sequestration" process helps balance the amount of carbon in the atmosphere, preventing the excess buildup of CO_2, which is a greenhouse gas responsible for global warming and climate change. By taking in CO_2 during photosynthesis, plants store carbon in their tissues, which can be transferred through the food web as animals consume plants, and eventually released back into the atmosphere through respiration and decomposition.

In addition to their role in the food chain, plants also act as carbon sinks. Forests, particularly tropical rainforests, and oceans are some of the largest carbon sinks on Earth. Phytoplankton in the oceans, for example, contribute significantly to global photosynthesis and carbon absorption. When plants die or shed leaves, the carbon stored in their biomass can be transferred to the soil, where it may remain for decades or centuries, especially in forests and wetlands. The oceans, too, absorb and store a large amount of carbon, with carbon being captured by marine plants and then either stored in the ocean or transported to the deep sea when these organisms die. However, human activities have disrupted the natural carbon cycle. Deforestation, land-use changes, and the burning of fossil fuels have significantly increased CO_2 concentrations in the atmosphere. As forests are cut down and degraded, the carbon stored in trees is released back into the atmosphere, and the Earth's ability to absorb and store CO_2 through photosynthesis is reduced. The burning of fossil fuels for energy releases large amounts of CO_2, further contributing to climate change. These human-induced changes intensify the greenhouse effect, raising global temperatures and creating a positive feedback loop, where higher temperatures can stress plant ecosystems, reducing their ability to photosynthesize and absorb CO_2. As global temperatures rise and climate change accelerates, the effects on photosynthesis could become more pronounced. Stressors such as heatwaves, droughts, and changing precipitation patterns can affect plant health and reduce their photosynthetic efficiency. This could, in turn, lower the capacity of natural carbon sinks to absorb CO_2, exacerbating the impacts of climate change. To mitigate these effects and help restore balance to the carbon cycle, efforts to protect and expand forests, improve agricultural practices, and enhance carbon sequestration in soils are essential. Reforestation, afforestation, and sustainable land management practices can increase the amount of CO_2 absorbed by plants and soils, while also improving ecosystem resilience. Additionally, technological solutions like carbon capture and storage (CCS) are being explored to help remove excess CO_2 from the atmosphere. Photosynthesis plays a crucial role in the global carbon cycle by removing CO_2 from the atmosphere and converting it into organic matter. This process helps maintain the balance of greenhouse gases, supports ecosystems, and mitigates the impacts of climate change. Preserving and enhancing photosynthetic processes, along with reducing human-induced CO_2 emissions, are key strategies for addressing the challenges posed by global warming and ensuring a sustainable future for the planet.

The Role of Plants in Mitigating Climate Change

Plants play a vital and multifaceted role in mitigating climate change by helping to regulate the Earth's climate system, reduce atmospheric greenhouse gases, and provide critical ecosystem services. Through processes like photosynthesis, carbon sequestration, and the support of biodiversity, plants act as both carbon sinks and sources of vital ecosystem functions that support life on Earth. Plants are fundamental allies in the fight against climate change. Through carbon sequestration, ecosystem services, and support for biodiversity, they help regulate the Earth's climate and create a more resilient environment. Protecting plant ecosystems, restoring forests, and implementing sustainable agricultural practices will not only mitigate climate change but also enhance the adaptive capacity of ecosystems and human communities alike. As we face an increasingly uncertain climate future, the role of plants in mitigating climate change has never been more important. Their role in climate change mitigation is increasingly recognized as essential in combating the adverse effects of global warming.

- ✰ **Carbon Sequestration and Storage:** One of the most significant ways plants help mitigate climate change is through carbon sequestration. During photosynthesis, plants absorb carbon dioxide (CO_2) from the atmosphere and convert it into glucose and other organic compounds. This carbon is then stored in the plant's tissues – in leaves, stems, and roots. As plants grow, they store carbon in biomass, which can remain locked away for extended periods, reducing the amount of CO_2 in the atmosphere.

- **Forests**: Forests, particularly tropical rainforests, are among the most effective carbon sinks on Earth. Trees in forests store large amounts of carbon, both in their biomass and in the soil. Forests act as critical reservoirs of carbon, absorbing more CO_2 than they release through respiration and decomposition.
- **Soils**: Plants also contribute to soil carbon storage. When plants shed leaves or die, organic material is deposited in the soil, where it can be stored for decades or even centuries. Healthy soils, rich in organic matter, act as significant carbon stores, helping to keep carbon out of the atmosphere.
- **Grasslands and Wetlands**: Grasslands and wetlands, although often overlooked, are also important carbon sinks. Wetlands, in particular, are capable of sequestering large amounts of carbon in waterlogged soils where decomposition is slow. Grasslands, which cover vast areas of the Earth, store substantial amounts of carbon in their root systems.

☆ **Reducing Greenhouse Gas Concentrations:** In addition to carbon sequestration, plants help mitigate climate change by reducing the overall concentration of greenhouse gases in the atmosphere. Through photosynthesis, plants absorb CO_2 – one of the most significant greenhouse gases contributing to global warming – and use it to fuel their growth. By acting as natural carbon sinks, forests, wetlands, and other ecosystems help maintain a balance of atmospheric CO_2, preventing its accumulation at harmful levels.

- **Oxygen Production**: The byproduct of photosynthesis is oxygen (O_2), which is released into the atmosphere. The oxygen produced by plants supports life on Earth, making it possible for animals and humans to breathe. The balance between oxygen and CO_2 is critical for the health of the planet's ecosystems.

☆ **Ecosystem Services for Climate Resilience:** Beyond their role in carbon sequestration, plants provide a range of ecosystem services that help mitigate the impacts of climate change and enhance the resilience of ecosystems. These services support human societies and natural systems alike:

- **Cooling the Environment**: Through a process called transpiration, plants release water vapor into the atmosphere, which helps cool the environment. Forests, urban green spaces, and agricultural lands with adequate vegetation help mitigate the "urban heat island" effect, where cities become significantly warmer than surrounding rural areas due to the concentration of heat-absorbing materials like concrete and asphalt.

- ❖ **Soil Erosion Prevention**: Plants, particularly grasses, shrubs, and trees, help prevent soil erosion. They bind the soil with their root systems, reducing the loss of soil carbon and preventing the release of greenhouse gases that would otherwise be trapped in the soil.
- ❖ **Water Regulation**: Vegetation plays an essential role in regulating the water cycle by absorbing rainwater and releasing it into the atmosphere. Plants in wetlands, forests, and grasslands help manage flood risks and mitigate droughts, which are expected to become more frequent and severe under climate change. Healthy plants and ecosystems can help buffer the impacts of extreme weather events.

☆ **Supporting Biodiversity:** Biodiversity is a key component of climate change resilience, and plants form the foundation of most ecosystems, supporting a wide variety of life. Healthy plant ecosystems provide food, shelter, and habitat for animals, insects, fungi, and microorganisms, all of which contribute to ecosystem stability. Biodiversity helps ecosystems function more effectively, enhancing their capacity to adapt to environmental changes, absorb carbon, and maintain other crucial ecosystem services.

- ❖ **Pollination and Crop Production**: Plants also support agriculture by providing pollinators, such as bees, butterflies, and birds, which are essential for the production of many food crops. Biodiversity in natural plant systems helps sustain the stability of food supplies, which is crucial as the impacts of climate change threaten global food security.

☆ **Plants in Climate Change Adaptation Strategies:** In addition to mitigation, plants are essential in adapting to the impacts of climate change. This includes reducing vulnerability to climate-related hazards such as flooding, drought, and heatwaves.

- ❖ **Afforestation and Reforestation**: Planting trees and restoring degraded forests (afforestation and reforestation) are effective ways to enhance carbon sequestration, restore biodiversity, and stabilize ecosystems. Reforestation efforts, especially in tropical and temperate zones, can significantly offset carbon emissions and improve the climate resilience of local communities.
- ❖ **Agroforestry and Sustainable Agriculture**: Integrating trees into agricultural systems (agroforestry) can enhance biodiversity, improve soil health, and increase carbon storage while providing farmers with additional income through timber and non-timber forest products.

Sustainable farming practices, such as no-till farming, crop rotation, and organic farming, can also help enhance soil carbon storage and reduce greenhouse gas emissions.

- **Challenges and the Need for Protection:** Despite the clear benefits of plants in mitigating climate change, their role is increasingly threatened by human activities. Deforestation, land-use changes, pollution, and climate-induced stresses, such as droughts and wildfires, are putting plants and ecosystems at risk. Deforestation alone is responsible for a significant portion of global CO_2 emissions, as it releases carbon stored in trees back into the atmosphere. To fully realize the potential of plants in mitigating climate change, it is essential to protect and restore ecosystems, implement sustainable land management practices, and promote policies that incentivize reforestation, afforestation, and the conservation of natural habitats.

The Carbon Footprint of Agriculture

Agriculture plays a significant role in the global carbon cycle and is a major contributor to the world's overall carbon footprint. The carbon footprint of agriculture refers to the total amount of greenhouse gases (GHGs) – primarily carbon dioxide (CO_2), methane (CH_4), and nitrous oxide (N_2O) – emitted directly or indirectly as a result of agricultural activities. These emissions come from various stages of food production, from land use changes and crop cultivation to livestock management and food processing. The carbon footprint of agriculture is a key issue in the context of climate change because agricultural practices not only emit GHGs but also have the potential to either exacerbate or mitigate climate change, depending on how they are managed.

Direct Emissions from Agricultural Activities

- **Soil Management and Fertilizer Use**: One of the primary contributors to the carbon footprint of agriculture is the use of synthetic fertilizers, particularly nitrogen-based fertilizers. When these fertilizers are applied to crops, they can lead to the release of nitrous oxide (N_2O), a potent greenhouse gas with a global warming potential approximately 298 times greater than CO_2 over a 100-year period. N_2O is released through soil processes such as nitrification and denitrification, especially when fertilizers are applied in excess or at inappropriate times. Poorly managed irrigation and over-fertilization also contribute to N_2O emissions.
- **Livestock Emissions (Methane)**: Livestock farming, particularly cattle and other ruminants, is another significant source of greenhouse gases, especially methane (CH). Methane is released during digestion (enteric fermentation) as part of the animals' metabolic processes. Cattle, sheep, and goats produce substantial amounts of methane, which is then released into the atmosphere through burping. Manure management also produces

methane, especially in systems where manure is stored in lagoons or pits, allowing anaerobic conditions that foster methane-producing bacteria.

- **Land Use and Deforestation**: Land-use change, especially the conversion of forests or wetlands into agricultural land, contributes significantly to the carbon footprint of agriculture. When forests are cleared for agriculture, the carbon stored in trees and vegetation is released into the atmosphere as CO_2. Deforestation for agriculture is responsible for around 10–15 per cent of global carbon emissions. In addition to CO_2 release, the destruction of ecosystems also reduces the capacity of forests to act as carbon sinks.
- **Energy Use in Agricultural Machinery**: Modern agricultural practices often rely heavily on fossil fuels for running machinery such as tractors, harvesters, irrigation pumps, and trucks for transportation. The combustion of diesel and gasoline for these machines produces CO_2 emissions. High-energy consumption in the processing and transport of food also contributes to the carbon footprint of agriculture.

Indirect Emissions from Agricultural Supply Chains

- **Food Processing and Transportation**: After crops are harvested, food products typically go through multiple stages of processing, packaging, and transportation. Each of these stages can add to the carbon footprint of food production. For example, meat production and processing are energy-intensive, requiring electricity and fossil fuels for refrigeration, transport, and packaging. The transportation of food from farms to markets, often involving long distances and refrigerated vehicles, adds further emissions. In the case of global trade, food may be transported by ships, trucks, or planes, all of which release significant amounts of CO_2.
- **Post-Harvest Losses and Wastage**: A significant portion of agricultural production is lost or wasted between harvest and consumption. According to the Food and Agriculture Organization (FAO), approximately one-third of food produced globally is lost or wasted, leading to unnecessary emissions. The carbon footprint associated with food waste includes the emissions generated from producing, processing, storing, and transporting food that is ultimately discarded. This not only represents inefficiency in food systems but also increases the overall carbon emissions associated with agriculture.

The Role of Agricultural Practices in the Carbon Footprint

- **Tillage and Soil Disturbance**: Conventional farming practices that involve plowing and tilling the soil can disturb carbon stored in the soil and release CO_2 into the atmosphere. Tilling accelerates the decomposition of organic matter in the soil, releasing stored carbon. The more frequently soil is tilled, the greater the potential for CO_2 release. Conversely, no-till or reduced-till farming practices can help reduce soil disturbance, enhance soil carbon sequestration, and improve soil health.

- **Irrigation**: While irrigation is crucial for crop growth, it can also indirectly increase greenhouse gas emissions. For example, irrigation systems that rely on fossil fuels to pump water or maintain pressure contribute to CO_2 emissions. Additionally, over-irrigation can lead to waterlogging and the release of methane from anaerobic soil conditions.

Mitigation Strategies to Reduce Agriculture's Carbon Footprint

There are several strategies that can help reduce the carbon footprint of agriculture and promote more sustainable practices:

- **Sustainable Fertilizer Use**: Efficient use of fertilizers, such as precision agriculture techniques, can minimize nitrous oxide emissions. By applying the right amount of fertilizer at the right time, and using organic or alternative fertilizers, emissions can be reduced. Integrated nutrient management, which combines organic and inorganic fertilizers, can also be a key strategy.
- **Reduced Livestock Emissions**: Reducing the carbon footprint of livestock farming involves changing animal diets, improving manure management practices, and exploring alternatives to traditional meat and dairy production. Adding feed additives that reduce methane production, such as fats or tannins, can help lower enteric methane emissions. Additionally, improving manure management by using aerobic processes or biogas systems can reduce methane emissions.
- **Agroforestry and Agroecological Practices**: Agroforestry – the integration of trees with crops and livestock – can help sequester carbon in both soil and vegetation. By planting trees on agricultural lands or restoring forests, farmers can enhance carbon storage and biodiversity. Agroecological practices, such as crop rotation, cover cropping, and organic farming, can improve soil health, increase carbon sequestration, and reduce the need for synthetic inputs like fertilizers and pesticides.
- **Soil Carbon Sequestration**: Practices such as no-till farming, cover cropping, and the use of organic matter (like compost) can enhance soil carbon sequestration, turning agricultural land into a carbon sink rather than a source of emissions. Carbon farming techniques, which focus on increasing carbon storage in soils, can be scaled up to reduce agriculture's overall carbon footprint.
- **Alternative Energy Sources**: Transitioning from fossil fuels to renewable energy sources for farm machinery, irrigation systems, and food processing can reduce the carbon emissions associated with agricultural

production. Solar, wind, and biogas technologies can all play a role in reducing agriculture's reliance on carbon-intensive energy.

- ☆ **Reducing Food Waste**: Addressing food waste at both the consumer and supply chain levels is another critical step in reducing the agricultural carbon footprint. By improving food storage, transportation, and consumption patterns, we can cut down on the emissions associated with lost and wasted food.

The Carbon Footprint of different Agricultural Products

Different types of food have vastly different carbon footprints. Animal products, particularly beef and dairy, are far more resource-intensive and carbon-intensive than plant-based foods.

- ☆ **Meat and Dairy**: Livestock farming is responsible for a substantial share of agricultural emissions. Beef, in particular, has one of the highest carbon footprints due to the methane produced by cattle during digestion, the need for large amounts of feed, and the energy-intensive nature of meat processing and transportation. Lamb and dairy also have relatively high carbon footprints, though not as high as beef.
- ☆ **Crops**: Crops like rice can also contribute significantly to the carbon footprint of agriculture, especially when grown in flooded fields, which can generate methane emissions. Other crops, like wheat, corn, and soy, tend to have lower carbon footprints per calorie or gram of protein compared to animal products.

The carbon footprint of agriculture is substantial, contributing significantly to global greenhouse gas emissions. Addressing this challenge requires a combination of changes in farming practices, dietary shifts, and innovations in technology. By adopting sustainable practices such as precision agriculture, better fertilizer management, reduced livestock emissions, and soil carbon sequestration, we can lower agriculture's carbon footprint. Additionally, reducing food waste and shifting to more plant-based diets can further decrease the overall environmental impact of food systems. As agriculture remains a cornerstone of global food security, its transformation into a more sustainable, low-carbon sector is crucial for achieving global climate goals and mitigating climate change.

Enhancing Photosynthesis for Food Security

Photosynthesis, the process by which plants convert sunlight into chemical energy, is the foundation of food production on Earth. It drives plant growth, supports the food chain, and contributes to the carbon cycle. Given the growing global population and the mounting challenges posed by climate change, enhancing photosynthesis to improve crop yields is a key strategy for ensuring food security in the future. By optimizing the efficiency of photosynthesis, we can increase the productivity of agricultural systems, improve the resilience of crops to environmental stressors, and ultimately feed more people with fewer resources.

The Role of Photosynthesis in Agriculture and Food Security

At its core, photosynthesis enables plants to convert carbon dioxide (CO_2), water, and sunlight into glucose ($C_6H_{12}O_6$), which serves as the energy source for growth and reproduction. This energy supports plant development, root expansion, and the production of seeds, fruits, and other edible parts of the plant. Since crops are the primary source of calories for humans and livestock, the efficiency of photosynthesis directly affects food availability, crop yields, and agricultural sustainability. In the context of food security, improving photosynthesis means boosting crop productivity while reducing the environmental impact of farming. The more efficiently crops photosynthesize, the more food can be produced from the same amount of land, water, and nutrients, which is essential as the global population continues to grow, and arable land becomes scarcer.

Challenges Facing Photosynthesis in Agriculture

Several factors currently limit the efficiency of photosynthesis in crops:

- **Climate Change**: Increasing temperatures, changing precipitation patterns, and more frequent extreme weather events such as droughts and heatwaves are negatively impacting photosynthetic efficiency. High temperatures, in particular, can cause plants to "shut down" photosynthesis or reduce its efficiency, leading to lower crop yields.
- **CO_2 Saturation**: While increasing atmospheric CO_2 might seem like a positive development for photosynthesis (since CO_2 is a key input in the process), studies show that the benefits are not as straightforward as expected. The increase in CO_2 concentration may stimulate photosynthesis in some crops, but it can also lead to nutrient imbalances in the plants, reducing the nutritional quality of the food produced (*e.g.*, lower protein content).
- **Photorespiration**: One of the key inefficiencies in photosynthesis is a process called photorespiration, which occurs in plants under high light and high temperature conditions. Photorespiration occurs when the enzyme RuBisCO, which is responsible for fixing CO_2, mistakenly binds to oxygen instead of carbon dioxide. This results in the wasteful use of energy and carbon, reducing the overall efficiency of photosynthesis. This issue is particularly prominent in crops such as rice, wheat, and soybeans, which are often grown in conditions that promote photorespiration.
- **Water Stress**: Water availability is another major factor that affects photosynthetic efficiency. In many regions, crops are subjected to water stress due to droughts, poor irrigation infrastructure, or competition for water resources. Water stress can cause stomatal closure (the pores on

leaves that allow for gas exchange), limiting the intake of CO_2 and reducing photosynthetic capacity.

Strategies to Enhance Photosynthesis for Food Security

To address these challenges and improve photosynthesis for the purpose of enhancing food security, several strategies have been proposed and are under research:

Genetic Engineering and Biotechnology

One of the most promising avenues for enhancing photosynthesis involves genetic engineering to create crops with improved photosynthetic efficiency. This can be achieved by targeting key genes involved in the photosynthetic process and modifying them to increase carbon fixation, reduce photorespiration, or improve the plant's ability to cope with environmental stress.

- ☆ **Improving RuBisCO$_2$ Efficiency**: RuBisCO$_2$ is the enzyme responsible for fixing CO_2 during the Calvin cycle, but it is relatively inefficient, especially under conditions of high temperature and low CO_2. Genetic modifications to improve the efficiency of RuBisCO$_2$ or to enhance its activity could significantly increase photosynthetic efficiency and yield potential. Some crops, such as rice and wheat, have been targeted for RuBisCO$_2$ improvements through genetic engineering.
- ☆ **Reducing Photorespiration**: Another target for genetic enhancement is the reduction of photorespiration. In some plants, researchers are developing methods to "shunt" photorespiratory pathways by introducing alternative metabolic pathways that reduce the losses associated with photorespiration. For instance, the development of "C rice" (which uses a more efficient photosynthetic pathway than the common C pathway) could lead to higher yields under high-temperature conditions.
- ☆ **Engineering C Photosynthesis in C Plants**: C photosynthesis, found in plants like maize (corn), sorghum, and sugarcane, is a more efficient photosynthetic pathway under conditions of heat and drought. Introducing genes that enable C crops (like rice, wheat, and soybean) to perform C photosynthesis could significantly boost productivity, particularly in regions affected by high temperatures and water scarcity.

Enhancing Crop Traits Through Breeding

Traditional plant breeding techniques, when combined with modern genomics, are also playing an important role in improving photosynthetic efficiency. Breeding crops that are better adapted to local climates, have deeper root systems, or are more efficient in water use can help enhance their photosynthetic performance.

- ☆ **Drought-Tolerant Crops**: Breeding crops for drought tolerance is crucial for maintaining food production in water-scarce regions. Crops that can maintain photosynthetic activity under reduced water availability will be better able to withstand climate change and continue to produce food in increasingly arid conditions.

- ☆ **High-Efficiency Crop Varieties**: Selecting and breeding varieties of crops that are better suited to specific climates or environments can optimize photosynthetic efficiency. This may include choosing varieties that have more extensive root systems for better nutrient and water uptake or those that are better able to maintain leaf function under stress.

Agronomic Practices

Beyond genetic modifications, agronomic practices can also be used to optimize photosynthesis and improve crop yields.

- ☆ **Precision Agriculture**: The use of technology such as sensors, drones, and satellite imaging allows farmers to monitor plant health and environmental conditions in real-time. This information can be used to optimize irrigation, fertilizer application, and planting practices, ensuring that crops receive the ideal conditions for photosynthesis.
- ☆ **Optimized Planting Densities**: Research has shown that adjusting planting densities can help maximize light capture and photosynthetic efficiency. By spacing crops appropriately, farmers can reduce competition for light, water, and nutrients, leading to better overall photosynthesis and higher yields.
- ☆ **Intercropping and Crop Rotation**: Growing multiple crops together (intercropping) or rotating different crops on the same land over time can help optimize resource use and improve photosynthetic efficiency. Some crops may help improve soil health, reduce pest pressure, or provide shade to reduce heat stress on other crops, all of which can indirectly enhance photosynthesis.

Climate-Resilient Farming Systems

Adapting farming systems to cope with climate change is essential to ensuring food security in the face of unpredictable weather patterns. Developing climate-resilient farming systems that support efficient photosynthesis involves:

- ☆ **Agroforestry**: Incorporating trees into farming systems can provide shade and wind protection for crops, helping to reduce heat stress and conserve water. Agroforestry also improves soil health, enhances carbon sequestration, and promotes biodiversity, all of which support sustainable agriculture.

Water Management

Efficient irrigation practices, such as drip irrigation and rainwater harvesting, help ensure that crops have the water they need to perform photosynthesis without depleting local water resources.

The Future of Photosynthesis in Food Security

Enhancing photosynthesis for food security is an exciting frontier in agricultural science. As climate change intensifies, the pressure to increase crop yields while reducing environmental impacts will grow. The combination of genetic engineering, precision agriculture, and sustainable farming practices offers a promising pathway to not only boost food production but also build resilience into food systems. In the coming years, breakthroughs in biotechnology – such as the development of crops with enhanced photosynthetic pathways, or even artificial photosynthesis systems – could dramatically increase agricultural productivity. However, these technologies must be carefully implemented to ensure that they are accessible, affordable, and equitable, particularly for smallholder farmers in developing countries who face the most significant challenges from climate change. Ultimately, enhancing photosynthesis is not just about increasing crop yields, but also ensuring that the food produced is nutritious, resilient to environmental stress, and sustainable over the long term. By improving the efficiency of photosynthesis, we can help secure a future where the growing global population has access to sufficient, nutritious food while mitigating the environmental impacts of agriculture.

Chapter 9

Metabolism of Secondary Metabolites

Introduction

Secondary metabolites are a diverse group of organic compounds produced by plants, fungi, bacteria, and other microorganisms that are not directly involved in the basic metabolic processes required for growth, development, or reproduction. Unlike primary metabolites, such as sugars, amino acids, and lipids, which are essential for cellular function, secondary metabolites often serve specialized functions related to an organism's survival, interaction with its environment, and defense mechanisms. These compounds include alkaloids, terpenoids, phenolics, flavonoids, and glycosides, among others. Their roles in nature are multifaceted; they can act as defenses against herbivores, pathogens, and competitors, as signaling molecules, or as agents involved in reproductive processes like attracting pollinators or seed dispersers. The metabolism of secondary metabolites is often more complex and tightly regulated compared to primary metabolites. These pathways are typically triggered by specific environmental factors or stress conditions, such as light, temperature, nutrient availability, or the presence of herbivores. The synthesis of secondary metabolites involves intricate biochemical routes, frequently initiated by primary metabolic intermediates, and often requires the action of specialized enzymes and regulatory networks. Many of these compounds are produced in low quantities, making their biosynthesis energetically costly for the organism, but the benefits they provide, such as protection from environmental stresses, enhanced competition, or increased reproductive success, make their production essential for survival in certain ecological niches. Understanding the metabolism of secondary metabolites is not only fundamental to the field of plant and microbial physiology but also holds significant practical applications in medicine, agriculture, and biotechnology. Many secondary metabolites possess pharmacological properties and have been used as sources of therapeutic agents, such as antibiotics, anticancer drugs, and pain relievers. Additionally, these metabolites are important in the food and beverage industry, as flavor compounds, preservatives, and colorants, and are being explored for their potential in sustainable agricultural practices and eco-friendly pest management. Thus, studying the metabolic pathways of secondary metabolites provides valuable insights into the complex biochemical interactions within organisms and offers potential avenues for novel applications in various industries.

Alkaloids, Terpenoids, and Phenolics

Alkaloids, Terpenoids, and Phenolics are three major classes of secondary metabolites that play key roles in the survival and ecological interactions of plants, fungi, bacteria, and other organisms. These compounds are often bioactive and have diverse functions ranging from defense mechanisms to communication with the environment. Alkaloids, terpenoids, and phenolics represent three key classes of secondary metabolites that play essential roles in plant defense, communication, and interaction with their environment. Alkaloids have powerful pharmacological effects, terpenoids are involved in aroma and defense, while phenolic compounds are primarily known for their antioxidant properties and contribution to plant pigmentation and protection. Together, these compounds not only contribute to the survival and reproductive success of the organisms that produce them, but they also hold significant value for human health and industry, particularly in medicine, agriculture, and food production.

Alkaloids

Alkaloids are a large group of nitrogen-containing compounds that are primarily synthesized by plants, although they are also found in fungi and some bacteria. Alkaloids are often derived from amino acids and are known for their potent pharmacological effects on humans and animals. The primary function of alkaloids in nature is believed to be defense against herbivores, pathogens, and competitors, as many alkaloids are toxic or deterrent to consumption. Some well-known alkaloids include morphine, caffeine, nicotine, quinine, and atropine.

Biosynthesis

The biosynthesis of alkaloids begins with the modification of amino acids, which serve as the primary precursors. For example, the amino acid tyrosine is a precursor to dopamine, which further leads to compounds like morphine in opium poppies. Alkaloids are generally synthesized in specific tissues (*e.g.*, roots, leaves, or seeds) and stored in vacuoles or other specialized structures. They are produced via complex biosynthetic pathways involving various enzymes, such as decarboxylases, methyltransferases, and reductases.

Types of Alkaloids

- **Indole Alkaloids**: Derived from tryptophan, examples include morphine, caffeine, and quinine.
- **Tropane Alkaloids**: Derived from ornithine, examples include atropine, scopolamine, and cocaine.
- **Purine Alkaloids**: Derived from purines, examples include caffeine, theobromine, and theophylline.
- **Steroidal Alkaloids**: These are derived from steroid precursors and include compounds like solanine and tomatine.

Pharmacological Effects

Alkaloids have profound effects on the nervous system, cardiovascular system, and metabolism. For example, nicotine is a stimulant, morphine is an analgesic, and quinine is an antimalarial agent. Many alkaloids are used in medicine and pharmaceuticals, either in their natural form or after modification to enhance their therapeutic effects.

Terpenoids (or Isoprenoids)

Terpenoids are a vast and diverse group of secondary metabolites that are derived from five-carbon isoprene units. They are the largest class of plant secondary metabolites and are essential in various biological processes, including growth, reproduction, and defense. Terpenoids play a critical role in interactions with the environment by deterring herbivores, attracting pollinators, or acting as antimicrobial agents. They are responsible for the aroma of many plants and are used in perfumes, flavorings, and traditional medicines.

Biosynthesis

Terpenoids are synthesized from isoprene (C5) units, which are polymerized to form larger molecules. The basic building blocks, isopentenyl pyrophosphate (IPP) and dimethylallyl pyrophosphate (DMAPP), are synthesized through two main pathways: the mevalonate pathway (found in the cytosol) and the methylerythritol phosphate (MEP) pathway (found in plastids). These pathways lead to the formation of different terpenoid skeletons, which are then modified by enzymes to create specific terpenoid compounds.

Types of Terpenoids

- **Monoterpenoids**: Composed of two isoprene units (C10), examples include menthol (from peppermint) and limonene (from citrus).
- **Sesquiterpenoids**: Composed of three isoprene units (C15), examples include farnesene and chamazulene.
- **Diterpenoids**: Composed of four isoprene units (C20), examples include gibberellins (plant hormones) and taxol (an anticancer drug).

- ☆ **Triterpenoids**: Composed of six isoprene units (C30), examples include squalene and betulinic acid.
- ☆ **Tetraterpenoids**: Composed of eight isoprene units (C40), examples include carotenoids like beta-carotene, which is a precursor to vitamin A.

Functions

- ☆ **Defense**: Many terpenoids act as toxins or deterrents to herbivores. For example, pyrethroids, derived from chrysanthemum flowers, are potent insecticides.
- ☆ **Aromatics**: Terpenoids contribute to the scent and flavor of many plants, such as the fragrant oils found in lavender, pine, and citrus.
- ☆ **Antimicrobial properties**: Many terpenoids, such as thymol and eucalyptol, exhibit antifungal and antibacterial activities.

Applications

Terpenoids have a wide range of uses in industry, including in the production of essential oils, fragrances, flavorings, and as bioactive compounds in traditional and modern medicine.

Phenolics

Phenolic compounds are a large group of secondary metabolites that contain one or more hydroxyl groups (-OH) attached to an aromatic ring. These compounds are widely distributed in the plant kingdom and play vital roles in plant growth, development, and defense. Phenolics are involved in protecting plants from UV radiation, pathogens, and herbivores, and they also contribute to the color, flavor, and aroma of many fruits, vegetables, and herbs.

Biosynthesis

Phenolics are synthesized through the shikimate pathway and the phenylpropanoid pathway, both of which originate from the amino acid phenylalanine. The phenylpropanoid pathway leads to the formation of a wide range of phenolic compounds, including flavonoids, lignans, tannins, and stilbenes. The shikimate pathway is involved in the production of essential aromatic amino acids, which are precursors for many phenolic compounds.

Types of Phenolic Compounds

- ☆ **Flavonoids**: These are one of the largest groups of phenolics, with a structure based on a 15-carbon skeleton. Examples include quercetin, kaempferol, and anthocyanins (which provide red, purple, and blue pigments in flowers and fruits).

- **Phenolic Acids**: Includes compounds like caffeic acid, ferulic acid, and chlorogenic acid, which are important in plant defense and have antioxidant properties.
- **Tannins**: High-molecular-weight polyphenolic compounds that can bind proteins and are involved in defense against herbivores. Tannins are found in tea, wine, and some fruits.
- **Lignans**: These are phenolic compounds that contribute to the structural integrity of plant cell walls, providing rigidity and resistance to herbivory.
- **Stilbenes**: These include resveratrol, found in grapes and wine, and are known for their antioxidant and anti-inflammatory properties.

Functions

- **Antioxidant Activity**: Many phenolics act as antioxidants, scavenging free radicals and protecting cells from oxidative damage. This property is particularly important in protecting plants from environmental stressors such as UV radiation and pathogen attack.
- **UV Protection**: Phenolic compounds, especially flavonoids, help protect plants from UV radiation by absorbing UV light and preventing DNA damage.
- **Defense against Herbivores and Pathogens**: Tannins and other phenolic compounds deter herbivory by making plant tissues less palatable or by directly inhibiting digestive enzymes in herbivores.
- **Color and Flavor**: Anthocyanins and other phenolic compounds contribute to the color and flavor of fruits, flowers, and other plant parts, aiding in pollination and seed dispersal.

Applications

Phenolic compounds are widely used in the food and beverage industry, particularly in wine, tea, and fruit products, due to their antioxidant and preservative properties. They are also studied for their potential therapeutic benefits, such as anti-inflammatory, anticancer, and heart-protective effects.

The Role of Secondary Metabolites in Plant Defence

Secondary metabolites, though not directly involved in the fundamental processes of growth and development, play crucial roles in a plant's ability to defend itself against various biotic and abiotic stressors. Plants are stationary organisms, and unlike animals, they cannot escape threats such as herbivory, pathogen infections, or environmental stress. To counteract these threats, plants produce a vast array of secondary metabolites that serve as defense compounds. These metabolites function in several ways to protect plants from herbivores, pathogens, and environmental damage, contributing to the plant's overall survival and reproductive success. The role of secondary metabolites in plant defense can be understood through their various mechanisms, which include deterrence, toxicity, antimicrobial action, and signaling.

Defense against Herbivores

One of the primary functions of secondary metabolites is to protect plants from herbivores, which may feed on their tissues, damage them, or affect their reproductive success. Plants have evolved a range of chemical defenses to deter herbivory, and these defenses are often sophisticated, involving direct toxicity, deterrence, or a combination of both.

Toxicity and Deterrence

- ☆ **Alkaloids**: Alkaloids such as nicotine, morphine, caffeine, and quinine are well-known for their toxicity. These compounds can interfere with the metabolism and nervous system of herbivores, often causing paralysis or death. For example, nicotine in tobacco acts as a potent neurotoxin for insects, and morphine in poppy plants can reduce feeding by herbivores.
- ☆ **Terpenoids**: Some terpenoids, like pyrethrins (from chrysanthemums), are potent insecticides. These compounds disrupt the nervous system of insects, effectively killing or repelling them. Monoterpenes like limonene and menthol also act as repellents for herbivores due to their strong odor, which is unpleasant to many pests.
- ☆ **Phenolics**: Tannins, a type of phenolic compound, bind to proteins and make plant tissues indigestible to herbivores. They reduce the nutritional quality of the plant, making it less palatable and more difficult to digest. Tannins also exhibit anti-digestive properties by inhibiting digestive enzymes in herbivores.

Mechanical Defense

In addition to chemical defenses, some secondary metabolites contribute to mechanical defense by influencing the structure of plant tissues. For example, lignans (a subclass of phenolic compounds) contribute to the lignification of plant cell walls, making them tougher and harder to chew, thus reducing the damage caused by herbivores.

Defense Against Pathogens

Secondary metabolites also play a vital role in protecting plants from bacterial, fungal, and viral infections. When pathogens invade plant tissues, they can cause diseases that lead to significant loss of plant vigor and productivity. Secondary metabolites help plants to defend against these invaders in several ways, including antimicrobial properties and inducing localized immune responses.

Antimicrobial Properties

- ☆ **Phenolic Compounds**: Many phenolic compounds, including flavonoids, tannins, and lignans, have been shown to possess antimicrobial properties. For example, tannins can inhibit the growth of fungi and bacteria by interfering with their cell wall synthesis or by precipitating proteins essential for microbial survival.

- **Terpenoids**: Some terpenoids, such as essential oils (*e.g.*, thymol from thyme, eucalyptol from eucalyptus), exhibit strong antifungal and antibacterial activities. These compounds are toxic to various pathogens and can prevent the spread of infections in plant tissues. For instance, eucalyptol has been shown to have antifungal properties against pathogens like *Fusarium* and *Phytophthora*.
- **Alkaloids**: Alkaloids like berberine and quinine have antimicrobial and antiprotozoal properties. In plants like *Berberis*, berberine has been observed to inhibit the growth of various bacterial and fungal pathogens.

Induced Resistance

In some cases, secondary metabolites are involved in induced resistance, where the plant's immune system is activated in response to a pathogen attack. This process often involves the production of defense-related secondary metabolites in response to stress signals:

- **Salicylic Acid**: A phenolic compound, salicylic acid, is central to the plant's systemic acquired resistance (SAR) response. It activates defense genes and leads to the accumulation of defensive compounds like pathogenesis-related proteins (PRs) and secondary metabolites that enhance pathogen resistance.
- **Jasmonic Acid**: Another plant hormone, jasmonic acid, triggers the biosynthesis of various terpenoids and alkaloids that have antimicrobial properties. It is part of the plant's wound response and is often produced when herbivores or pathogens damage the plant tissue.

Defense Against Environmental Stressors

In addition to biotic threats, secondary metabolites help plants cope with abiotic stresses such as UV radiation, extreme temperatures, drought, and pollution. These environmental stressors can damage plant tissues and reduce overall fitness, but secondary metabolites often act as protectants in such situations.

UV Protection

- **Flavonoids**: Flavonoids, particularly anthocyanins and quercetin, absorb ultraviolet (UV) light and protect plant tissues from UV-induced oxidative damage. This UV-absorbing property helps to shield sensitive cellular structures, such as DNA and chloroplasts, from harmful radiation.
- **Phenolic Acids**: Compounds like caffeic acid and ferulic acid also contribute to UV protection by absorbing UV radiation and acting as antioxidants, reducing oxidative stress caused by UV exposure.

Antioxidant Defense

- **Phenolic Compounds**: Many phenolic compounds act as antioxidants, neutralizing reactive oxygen species (ROS) that are produced during oxidative stress. For example, flavonoids and tannins scavenge free radicals generated by UV radiation or pathogen attack, thus protecting the plant from cellular damage.
- **Terpenoids**: Certain terpenoids also possess antioxidant properties. For example, carotenoids (such as beta-carotene) protect against oxidative stress caused by light and environmental pollutants by scavenging free radicals.

Drought and Water Stress

- **Terpenoids and Phenolic Compounds**: Many terpenoids and phenolics are involved in osmoregulation, helping plants cope with water stress. For example, terpenoids like abscisic acid (ABA) play a role in regulating stomatal closure, reducing water loss during drought conditions. Phenolics, including flavonoids, can also help by stabilizing cell membranes and proteins, minimizing damage during desiccation.

Signaling and Communication

Secondary metabolites are involved in plant communication, particularly in signaling defense responses to other plants, herbivores, or pathogens. When a plant is attacked by herbivores or pathogens, it can produce volatile organic compounds (VOCs) that serve as signals to nearby plants or other organisms.

- **Volatile Terpenoids**: These are often released in response to herbivore damage or pathogen infection. For example, when attacked by herbivores, plants like *tobacco* release VOCs such as *limonene* and *myrcene*, which can attract natural enemies of herbivores (*e.g.*, parasitoid wasps) to help control the herbivore population.
- **Systemic Signaling**: Secondary metabolites, like jasmonic acid and salicylic acid, are involved in long-distance signaling within the plant. After an initial attack, these compounds help the plant activate defense mechanisms in uninjured tissues, providing enhanced resistance to further damage.

Role of Secondary Metabolites in Coevolution

Plants and their herbivores, pathogens, and pollinators are involved in a coevolutionary "arms race." As herbivores and pathogens evolve mechanisms to overcome plant defenses, plants, in turn, evolve more sophisticated secondary metabolites to counteract these threats. This dynamic process leads to an ongoing evolution of both plant and herbivore/predator strategies. For example:

- ☆ **Herbivore Resistance**: Some herbivores evolve the ability to detoxify or tolerate plant secondary metabolites. In response, plants may produce even more potent toxins or diversify the types of secondary metabolites they produce.
- ☆ **Pathogen Adaptation**: Similarly, pathogens may evolve resistance mechanisms to plant-produced antimicrobial compounds, prompting plants to evolve new types of secondary metabolites or enhance the concentration of existing ones.

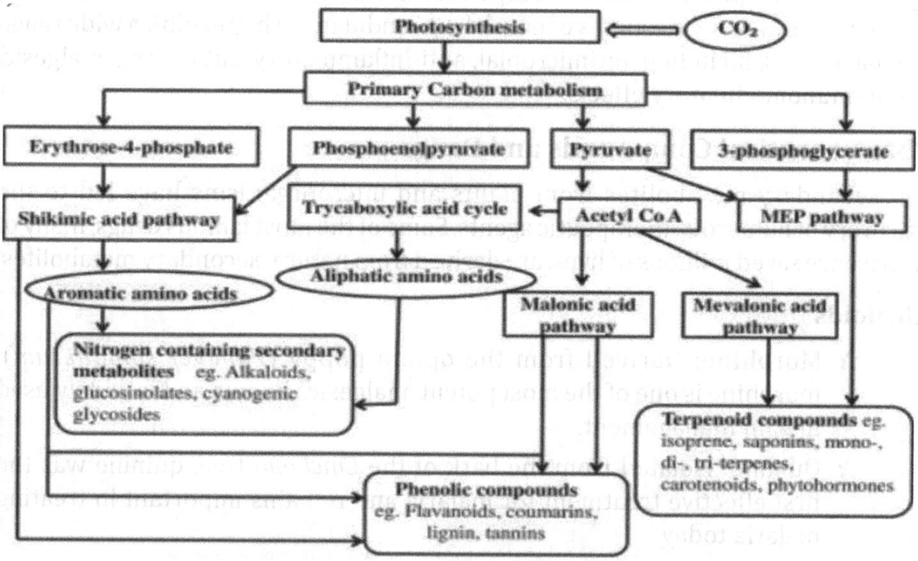

Figure 9.1. General Pathways of Formation of Secondary Metabolites.

Secondary metabolites play an essential role in the defense mechanisms of plants. From deterring herbivores to defending against pathogens and environmental stresses, these compounds are vital for plant survival. Their diverse range of functions–ranging from direct toxicity and antimicrobial activity to signaling and communication–demonstrates the complexity and versatility of plant defense systems. Understanding how these metabolites are synthesized, regulated, and utilized can lead to advancements in agriculture, pest management, and even the development of new pharmaceuticals. In this way, secondary metabolites not only contribute to the survival of plants but also offer valuable benefits to humans.

Applications in Medicine, Agriculture and Biotechnology

Secondary metabolites, though not essential for the basic growth and development of plants, play pivotal roles in their interaction with the environment and contribute significantly to their survival. Beyond their ecological functions, secondary metabolites are also valuable to humans in various fields, including medicine, agriculture, and biotechnology. These compounds have diverse applications, from serving as raw materials for pharmaceutical drugs and agricultural products to enabling biotechnological innovations aimed at improving health, food production, and sustainability. Let's explore in detail how secondary metabolites are utilized in these key areas.

Applications in Medicine

Secondary metabolites have long been a cornerstone of traditional and modern medicine. Many pharmaceutical compounds are derived from these metabolites, and they continue to be sources of novel drug candidates. They exhibit a wide range of bioactivities, including antimicrobial, anti-inflammatory, anticancer, analgesic, and immunomodulatory effects.

Pharmaceutical Compounds and Drugs

Secondary metabolites from plants and microorganisms have led to the discovery of numerous therapeutic agents. Some of the most famous drugs, many of which have saved millions of lives, are derived from natural secondary metabolites.

Alkaloids

- Morphine: Derived from the opium poppy (*Papaver somniferum*), morphine is one of the most potent analgesics known and is widely used in pain management.
- Quinine: Isolated from the bark of the *Cinchona* tree, quinine was the first effective treatment for malaria and remains important in treating malaria today.
- Caffeine: Found in coffee, tea, and various other plants, caffeine is a stimulant that is used in medications for headache relief and as a cognitive enhancer.
- Nicotine: Though harmful when abused, nicotine from tobacco has been studied for its potential in treating neurodegenerative diseases such as Parkinson's and Alzheimer's due to its effects on neurotransmitter systems.

Terpenoids

- **Taxol (Paclitaxel):** This diterpenoid, derived from the bark of the Pacific yew tree (*Taxus brevifolia*), is a highly effective anticancer drug, particularly in the treatment of ovarian, breast, and lung cancers.
- **Artemisinin:** Extracted from the sweet wormwood plant (*Artemisia annua*), artemisinin is a powerful antimalarial compound, and its derivatives are among the most effective treatments for malaria.

Phenolic Compounds

- **Resveratrol:** A stilbene phenolic compound found in grapes, resveratrol has garnered significant attention for its potential cardiovascular benefits, antioxidant properties, and possible anticancer effects.
- **Flavonoids:** Many flavonoids, such as quercetin and catechins (found in apples, onions, and green tea), have potent antioxidant and anti-inflammatory activities and are studied for their potential in preventing chronic diseases like cardiovascular disease, cancer, and diabetes.

Antibiotics and Antifungals

Secondary metabolites from microorganisms have been the source of many essential antibiotics and antifungal agents:

- **Penicillin:** Derived from the fungus *Penicillium chrysogenum*, penicillin was the first antibiotic discovered and remains in use today to treat bacterial infections.
- **Streptomycin:** Isolated from *Streptomyces griseus*, this antibiotic is particularly effective against tuberculosis and other bacterial diseases.

Vaccine Development and Immunotherapy

Some secondary metabolites are involved in the development of vaccines and immunotherapies. For example, certain plant-derived compounds are used as adjuvants in vaccines to enhance the body's immune response.

Applications in Agriculture

Secondary metabolites in plants have profound implications for agriculture, particularly in terms of crop protection, pest management, and improving crop quality. These compounds can provide natural alternatives to synthetic pesticides and herbicides, and also contribute to the development of crops with better nutritional profiles and resistance to stress.

Biopesticides and Plant Protection

Natural products derived from secondary metabolites are used as biopesticides in integrated pest management (IPM). These biopesticides are often less toxic to humans, animals, and beneficial insects compared to synthetic chemicals, making them more environmentally friendly.

- **Pyrethrins:** Derived from chrysanthemums, pyrethrins are potent insecticides that disrupt the nervous system of insects. They are widely used in agriculture and are often found in natural insect repellents and pet shampoos.
- **Neem Oil:** Derived from the seeds of the neem tree (*Azadirachta indica*), neem oil contains a variety of terpenoid compounds, such as azadirachtin, that have insecticidal, antifungal, and antibacterial properties.

- ☆ **Alkaloids:** Certain alkaloids, such as those from the *Tobacco* plant, have been shown to possess insecticidal and herbicidal properties, and they are used in controlling pest populations.

Herbicide Alternatives

Natural herbicides derived from plant secondary metabolites are being developed as alternatives to synthetic herbicides, which are often harmful to the environment. For example:

- ☆ Allelopathy: Some plants produce secondary metabolites that inhibit the germination or growth of competing plants. These compounds are studied for their potential use in natural herbicide formulations.
- ☆ Glucosinolates: Found in plants of the Brassicaceae family (*e.g.*, mustard, cabbage), glucosinolates and their breakdown products have shown promise as natural herbicides.

Improving Crop Resistance

Plants produce a variety of secondary metabolites in response to biotic (*e.g.*, pests, pathogens) and abiotic (*e.g.*, drought, temperature extremes) stresses. These metabolites contribute to the plant's ability to withstand these stresses, and their manipulation can lead to improved crop varieties:

- ☆ **Flavonoids and Phenolic Compounds:** These compounds play a role in plant defense against pathogens and herbivores, as well as in mitigating the effects of UV radiation and oxidative stress.
- ☆ **Terpenoids:** Some terpenoids help plants cope with environmental stress, such as drought or high salinity, by functioning as antioxidants or by modifying the plant's hormonal responses.

By understanding and enhancing the production of these metabolites in crops, agriculturalists can breed plants that are more resistant to pests, diseases, and environmental stressors, leading to improved yields and reduced reliance on chemical inputs.

Nutritional Enhancement of Crops

Secondary metabolites can also contribute to improving the nutritional content of crops. For example, enhancing the levels of carotenoids (such as beta-carotene) in crops like rice (Golden Rice) provides increased levels of provitamin A, addressing nutrient deficiencies in developing regions.

Applications in Biotechnology

Biotechnology offers opportunities to harness the power of secondary metabolites for industrial applications. The genetic and enzymatic pathways responsible for the synthesis of secondary metabolites can be optimized, and microorganisms or plants can be engineered to produce these valuable compounds in large quantities. Secondary metabolites are invaluable in a wide range of applications that span medicine, agriculture, and biotechnology. From life-saving drugs to eco-friendly pesticides, these compounds have played a key role in improving human health, agricultural practices, and industrial processes. As research continues, the potential to engineer organisms to produce these compounds in a sustainable and cost-effective manner offers exciting opportunities for developing new solutions to global challenges, including the treatment of diseases, enhancement of crop productivity, and the creation of sustainable bio-based products. Understanding the biosynthesis, regulation, and application of secondary metabolites will continue to be an important focus for researchers in these fields.

Metabolic Engineering

Metabolic engineering is a powerful tool that allows scientists to manipulate the biosynthetic pathways of secondary metabolites, improving their yields or enabling the production of compounds that are difficult to obtain from natural sources.

- ✰ **Yeast and Bacterial Systems:** Microorganisms like *Saccharomyces cerevisiae* (yeast) and *Escherichia coli* are used in industrial biotechnology to produce secondary metabolites like artemisinin (used in malaria treatment) through engineered biosynthetic pathways. This approach provides a more cost-effective and sustainable method for producing high-value compounds.
- ✰ **Plant Cell Cultures:** In vitro plant cell culture systems are being developed to produce secondary metabolites such as alkaloids, flavonoids, and terpenoids. These cultures can be optimized to produce large quantities of desired compounds, reducing the reliance on natural plant sources and minimizing environmental impact.

Biotransformation

Biotransformation refers to the use of microorganisms or enzymes to modify natural compounds into more valuable or bioactive forms. For example, microorganisms can convert simple terpenoid precursors into more complex and pharmacologically active molecules. This process is useful in the production of pharmaceutical compounds or in generating novel derivatives with enhanced properties.

Synthetic Biology

Synthetic biology combines principles from molecular biology, engineering, and systems biology to design new, artificial biosynthetic pathways for secondary metabolites. This allows the production of novel compounds that do not naturally occur in nature, potentially leading to new drugs, biofuels, or other industrial products. For example, synthetic biology is being explored to create biofuels derived from terpenoid compounds or to engineer microorganisms that can produce high-value secondary metabolites like resveratrol or paclitaxel on an industrial scale.

Environmental Applications

Secondary metabolites produced by microorganisms are also being explored for bioremediation, or the use of living organisms to clean up environmental pollutants. For example, certain fungi and bacteria produce secondary metabolites that can degrade or transform toxic substances, such as heavy metals or industrial chemicals, into less harmful forms.

Chapter 10

Plant Biotechnology and Genetic Engineering

Introduction

Plant biotechnology and genetic engineering are transformative fields at the intersection of biology, technology, and agriculture, offering innovative solutions to some of the world's most pressing challenges. These disciplines leverage cutting-edge techniques to modify the genetic makeup of plants, enabling scientists to introduce desirable traits that would be difficult or impossible to achieve through traditional breeding methods. By harnessing the power of molecular biology, gene editing, and tissue culture, plant biotechnology has revolutionized the way we grow, protect, and utilize crops, leading to enhanced yields, improved resistance to diseases and pests, and the development of plants with novel nutritional profiles. Genetic engineering, in particular, has opened up new possibilities for creating genetically modified organisms (GMOs) that can thrive in a variety of environmental conditions, from drought-prone regions to areas affected by soil salinity. These advances have significant implications for global food security, environmental sustainability, and climate change mitigation, offering a promising pathway towards more resilient agricultural systems. Despite the tremendous potential, the fields of plant biotechnology and genetic engineering also raise important ethical, environmental, and regulatory questions, making the responsible application of these technologies a key area of ongoing research and debate.

Genetic Engineering in Plants: Techniques and Tools

Genetic engineering in plants is a process that involves the manipulation of a plant's DNA to introduce new traits, enhance existing characteristics, or develop entirely novel plant varieties. This is achieved using a variety of techniques and tools that allow for the precise insertion, deletion, or modification of specific genes within the plant genome. The goal of genetic engineering is to produce plants with improved qualities, such as enhanced resistance to pests, diseases, and environmental stresses, increased nutritional content, or better yield potential. The tools and techniques used in plant genetic engineering have significantly advanced our ability to manipulate plant genomes with precision and efficiency. From the use of Agrobacterium and gene guns for transformation to the development of sophisticated gene-editing systems like CRISPR/Cas9, these methods enable scientists to create crops with enhanced traits that can help address global challenges, such as food security, climate change, and environmental sustainability. As the field progresses, new technologies and applications will continue to emerge, further expanding the potential of genetic engineering to revolutionize agriculture and plant biotechnology.

Several advanced tools and methods are used to achieve these genetic modifications, including the following:

Gene Cloning and Transformation

The first step in genetic engineering involves isolating and cloning the gene of interest. This gene may come from the same plant species (cisgenic), a closely related species (intragenic), or a completely different organism (transgenic). Gene cloning is achieved using several methods, including:

- **Plasmid Vectors:** Plasmids, which are small circular DNA molecules found in bacteria, are commonly used as vectors to carry the gene of interest into plant cells. A gene is first inserted into the plasmid using restriction enzymes and ligases, creating a recombinant DNA molecule. This plasmid is then introduced into plant cells via one of several transformation techniques.
- **Agrobacterium-mediated Transformation:** The bacterium *Agrobacterium tumefaciens* is a natural genetic engineer, capable of transferring a segment of its own DNA (the T-DNA) into the genomes of plant cells. Scientists have harnessed this natural mechanism by replacing the T-DNA with a gene of interest. The modified *Agrobacterium* is then used to infect plant cells, transferring the new gene into the plant's genome. This method is particularly effective for dicot plants (such as tomatoes, soybeans, and cotton).
- **Gene Gun (Biolistics):** The gene gun, or biolistic particle delivery system, is a technique used to deliver DNA directly into plant cells. Tiny gold or tungsten particles are coated with the gene of interest and shot at high velocity into plant tissues. This method is particularly useful for monocots (such as corn and wheat) and can be used to transform plant cells that are difficult to infect with *Agrobacterium*.

Gene Editing Techniques

Gene editing has revolutionized the field of plant biotechnology, enabling precise modifications to the plant genome without the need to introduce foreign DNA. The most commonly used gene-editing technologies include:

- ☆ **CRISPR/Cas9:** The CRISPR (Clustered Regularly Interspaced Short Palindromic Repeats) system, coupled with the Cas9 enzyme, is a groundbreaking gene-editing tool that allows for precise, targeted modifications to the plant genome. The system works by creating a double-strand break at a specific location in the DNA, which the cell then attempts to repair. During this repair process, the DNA sequence can be altered by introducing a desired change. This method is highly versatile, enabling the deletion, insertion, or modification of specific genes, with applications ranging from improving disease resistance to enhancing drought tolerance.
- ☆ **TALENs (Transcription Activator-Like Effector Nucleases):** TALENs are another type of gene-editing tool that work by recognizing specific DNA sequences and inducing double-strand breaks at those locations. TALENs are made by fusing a DNA-binding domain (the TAL effector) with a nuclease domain (which cuts the DNA). Like CRISPR, TALENs can be used to insert, delete, or modify specific genes, though the technology is generally more complex and less widely used compared to CRISPR.
- ☆ **ZFNs (Zinc Finger Nucleases):** Zinc Finger Nucleases are another class of gene-editing tools that use zinc finger proteins to recognize and bind to specific DNA sequences. When combined with a nuclease domain, they can create targeted double-strand breaks in the DNA. While ZFNs were one of the first gene-editing tools developed, they are less commonly used today due to the complexity of designing specific zinc finger proteins.

Somatic Hybridization and Protoplast Fusion

Somatic hybridization involves the fusion of protoplasts–plant cells that have had their cell walls removed–allowing for the mixing of the genomes of different plant species or varieties. This method can be used to combine traits from two genetically distinct plants that would not normally cross-breed. After fusion, the hybrid cells are regenerated into full plants. This technique is particularly useful for creating hybrids between species that are genetically distant or incompatible through traditional breeding methods.

RNA Interference (RNAi)

RNA interference is a technique used to silence specific genes in a plant. This method works by introducing double-stranded RNA (dsRNA) molecules that are complementary to the target gene's mRNA. The RNAi machinery in the plant then degrades the mRNA, preventing the gene from being expressed. RNAi has been used to create plants with improved resistance to pests and diseases, and to reduce the levels of harmful compounds, such as allergens or toxins, in food crops.

CRISPR/Cas12a and Base Editing

An extension of the CRISPR/Cas9 system, CRISPR/Cas12a (also known as Cpf1), is a newer tool for gene editing that offers greater precision and fewer off-target effects. It differs from Cas9 in terms of its cutting mechanism and the type of DNA sequences it recognizes. Base editing, a modification of the CRISPR/Cas system, allows for the direct conversion of one DNA base pair into another without creating double-strand breaks, offering a safer and more efficient method for making small, precise changes in the plant genome.

Plant Transformation via Agrobacterium and CRISPR

A novel approach that combines both the Agrobacterium method and CRISPR/Cas9 technology has emerged, where CRISPR components are directly delivered into the plant cells via Agrobacterium-mediated transformation. This approach allows for the efficient editing of plant genomes with high precision and reduced off-target effects, combining the strengths of both tools.

Molecular Markers and Marker-Assisted Selection (MAS)

Molecular markers are specific DNA sequences associated with desirable traits, such as disease resistance or high yield. These markers are used in Marker-Assisted Selection (MAS) to select plants that possess the desired traits without the need for time-consuming phenotypic evaluation. In genetic engineering, molecular markers are often used to confirm the successful incorporation of the gene of interest into the plant genome, aiding in the screening and selection of transformed plants.

Plant Tissue Culture and Regeneration

Plant tissue culture techniques, including micropropagation and organogenesis, are often used to regenerate whole plants from single plant cells or tissues that have been genetically modified. The plant tissue is cultured on nutrient media that supports cell division and growth, eventually forming shoots, roots, and complete plants. This is a crucial step in the genetic engineering process, as it enables the propagation of genetically modified plants for field trials, commercial production, or further research.

Biotechnology in Precision Agriculture

Genetic engineering tools, particularly CRISPR and gene cloning techniques, play a central role in the development of precision agriculture, a technology-driven approach to farming that optimizes crop yields while minimizing environmental impact. These tools are used to create crops that are resistant to specific pests, diseases, or environmental stresses, or that have improved nutritional profiles. As part of precision agriculture, genetically engineered plants are often used in conjunction with technologies like drone monitoring, soil sensors, and automated irrigation systems to ensure optimal growth conditions and reduce resource waste.

Transgenic Plants and their Applications

Transgenic plants are plants that have been genetically modified (GM) to contain one or more foreign genes–genes from other species or varieties–that impart desirable traits not naturally found in the plant. These plants are developed through genetic engineering techniques that allow scientists to insert specific genes into the plant genome. These genes can come from other plants, microorganisms, animals, or even synthetic sources. The resulting genetically modified (GM) plants, often referred to as transgenic plants, can exhibit improved characteristics such as higher resistance to pests, tolerance to environmental stresses, enhanced nutritional content, or the production of valuable compounds. The creation of transgenic plants is a key development in the field of agricultural biotechnology and has significant applications in food production, environmental sustainability, industrial uses, and medicine. The process typically involves inserting a gene of interest into the plant's genome using techniques such as Agrobacterium-mediated transformation, gene gun, or CRISPR/Cas9 gene editing. Once the foreign gene is successfully integrated into the plant genome, the plant expresses the desired trait, which can be passed on to future generations through conventional breeding or seed production.

Applications of Transgenic Plants

The applications of transgenic plants span a wide range of areas, from agriculture to industrial biotechnology and pharmaceuticals. Transgenic plants hold great promise for addressing global challenges such as food security, climate change, and health issues. Through genetic modification, crops can be made more resistant to pests, diseases, and environmental stresses, while also improving nutritional content and supporting industrial applications like biofuels and pharmaceuticals. However, the widespread adoption of these plants requires careful regulation and ongoing scientific research to mitigate potential risks to human health, the environment, and global agricultural systems. With proper oversight and continued innovation, transgenic plants could play a significant role in shaping the future of agriculture and biotechnology.

Here are some of the most significant applications:

Agricultural Improvement

- ☆ **Pest Resistance:** One of the earliest and most successful applications of transgenic plants has been the development of crops that are resistant to insect pests. A well-known example is Bt cotton, which has been genetically engineered to express the toxin from *Bacillus thuringiensis* (Bt), a bacterium that kills specific insect pests, particularly the cotton bollworm. This has led to reduced pesticide use, lower production costs, and increased crop yields. Other transgenic crops, such as Bt corn and Bt rice, also contain genes from *Bacillus thuringiensis* to confer resistance to pests like the European corn borer, reducing the need for chemical insecticides.

- ☆ **Herbicide Tolerance:** Another major application of transgenic plants is the development of crops that can tolerate specific herbicides, which makes weed control easier and more efficient. Glyphosate-tolerant crops, such as Roundup Ready soybeans, corn, and cotton, are engineered to withstand the herbicide glyphosate, allowing farmers to use the herbicide to control weeds without harming the crop itself. This has simplified weed management in large-scale farming and helped reduce the reliance on more toxic or labor-intensive weed control methods.
- ☆ **Disease Resistance:** Transgenic plants can also be engineered for resistance to viral, fungal, and bacterial diseases. For example, Papaya ringspot virus-resistant papaya is a widely used transgenic variety that saved papaya crops in Hawaii from the devastating effects of the ringspot virus. Tomato plants have also been modified to resist the tomato mosaic virus and late blight, the pathogen responsible for the Irish potato famine.
- ☆ **Drought and Stress Tolerance:** With climate change leading to more frequent and severe droughts, transgenic plants that can tolerate water scarcity have become crucial. Plants such as drought-resistant corn and rice have been engineered to express genes that help them cope with reduced water availability by improving water retention or altering the plant's metabolic pathways. Similarly, salt-tolerant transgenic plants, such as salt-tolerant tomatoes and rice, are being developed to thrive in saline soils, expanding the areas where crops can be grown.
- ☆ **Improved Yield:** Genetic modification can also increase crop yields by enhancing photosynthesis, improving nutrient uptake, or boosting plant growth. For example, transgenic golden rice, which contains added genes for the production of pro-vitamin A (beta-carotene), addresses vitamin A deficiency in populations dependent on rice as a staple food.

Nutritional Enhancement

- ☆ **Golden Rice:** Golden rice is a notable example of a transgenic plant designed to address nutritional deficiencies, particularly vitamin A deficiency. The rice is engineered to produce higher levels of beta-carotene, which the human body can convert into vitamin A. This genetically modified rice could play a critical role in combating malnutrition in developing countries, where rice is a staple food but the diet often lacks sufficient vitamin A.
- ☆ **Enhanced Protein and Mineral Content:** Genetic engineering has also been used to enhance the nutritional profile of plants by increasing the

levels of essential amino acids, proteins, and minerals. For example, soybeans have been modified to produce higher levels of specific amino acids such as methionine, which is particularly important for livestock feed. Similarly, biofortified maize with higher levels of lysine and tryptophan has been developed to improve the nutritional quality of maize, especially in regions where this grain is a dietary staple.

Environmental Benefits

- **Reduced Pesticide Use:** Transgenic plants that are resistant to pests (*e.g.*, Bt cotton, Bt corn) can reduce the need for chemical pesticides, which not only lowers production costs but also decreases the environmental impact of pesticide use. This leads to better soil and water quality, reduced harm to non-target species (including beneficial insects like bees), and a reduction in pesticide resistance.
- **Soil Health and Conservation:** Herbicide-tolerant crops, such as Roundup Ready crops, allow for the adoption of conservation tillage practices, where fewer passes with heavy machinery are needed to control weeds. This leads to less soil erosion, better water retention, and overall improved soil health.

Industrial Applications

- **Bioremediation:** Transgenic plants have been developed to clean up contaminated soils and waters through a process known as bioremediation. These plants can be engineered to absorb and detoxify pollutants such as heavy metals (*e.g.*, lead, mercury), pesticides, or petroleum products. Transgenic poplar trees and mustard plants have been developed for this purpose, capable of taking up and storing heavy metals from contaminated environments.
- **Pharmaceutical Production:** Transgenic plants are also used in the production of pharmaceuticals, a field known as pharming. Plants can be engineered to produce proteins, vaccines, and other therapeutic compounds. For example, tobacco plants have been modified to produce proteins used in the treatment of diseases like cystic fibrosis and hemophilia. Additionally, banana plants have been engineered to produce vaccines against diseases such as hepatitis B and diarrheal diseases.
- **Bioplastics and Biofuels:** Certain transgenic plants are engineered to produce materials like bioplastics or biofuels, which offer environmentally friendly alternatives to petroleum-based products. Oilseed crops like soybeans and camelina have been genetically modified to produce higher quantities of oils that can be converted into biodiesel. Similarly, plants such as switchgrass are being modified for use as a sustainable biofuel source, with reduced competition with food production.

Pharmaceutical and Medical Applications

- **Production of Therapeutic Proteins and Vaccines:** Transgenic plants have been used to produce therapeutic proteins for the treatment of various diseases. For example, genetically engineered tobacco plants have been used to produce human antibodies and enzymes for pharmaceutical applications. The use of plants as "biofactories" allows for the production of proteins that would be expensive or difficult to produce in traditional systems like bacterial or mammalian cell cultures.
- **Edible Vaccines:** Another exciting application of transgenic plants is the development of edible vaccines, where genetically modified plants produce antigens that can trigger an immune response in humans. For instance, potatoes and tomatoes have been genetically engineered to produce antigens against diseases like cholera and hepatitis B, potentially providing a low-cost, easily administered form of vaccination, particularly in developing countries.

Challenges and Controversies

While the potential benefits of transgenic plants are immense, their use has been met with various challenges and controversies. Some of the key concerns include:

- **Environmental Risks:** There are concerns about the unintended spread of transgenic traits to wild relatives or non-GM crops (gene flow), potentially creating "superweeds" or other ecological imbalances.
- **Health and Safety:** Public concerns about the safety of genetically modified foods persist, especially regarding potential allergic reactions, long-term health effects, and the impact on biodiversity.
- **Ethical and Socioeconomic Issues:** The monopolization of genetic technologies by a few large biotech companies has raised concerns about the control of seed markets, the displacement of smallholder farmers, and the ethical considerations of altering the genetic makeup of living organisms.

Genetically Modified Crops: Benefits and Controversies

Genetically modified (GM) crops, also known as genetically engineered (GE) crops, are plants that have been altered through genetic engineering techniques to exhibit desirable traits that are not naturally found in the species. These modifications can include resistance to pests and diseases, tolerance to environmental stresses (such as drought or salinity), and improvements in nutritional content or growth rates. GM crops are the result of inserting specific genes into a plant's genome, often from other species, through techniques like gene transfer, gene editing (*e.g.*, CRISPR), and Agrobacterium-mediated transformation. While genetically modified crops have been widely adopted in many parts of the world, particularly in the United States, Brazil, and Argentina, their use remains highly controversial. Proponents argue that these crops offer a range of benefits for food security, sustainability, and economic growth, while opponents raise concerns about environmental, health, and ethical risks. Here, we will explore both the benefits and controversies surrounding GM crops.

Benefits of Genetically Modified Crops

Increased Agricultural Productivity

One of the most significant advantages of genetically modified crops is their ability to increase agricultural productivity. By incorporating traits such as pest resistance, disease resistance, and herbicide tolerance, GM crops help farmers achieve higher yields with fewer inputs.

- ☆ **Pest Resistance:** Crops like Bt cotton and Bt corn are genetically modified to express proteins from the bacterium *Bacillus thuringiensis*, which act as natural insecticides. These crops are resistant to pests such as the cotton bollworm and European corn borer, reducing the need for chemical pesticides. This leads to higher yields, reduced pesticide use, and lower production costs.
- ☆ **Disease Resistance:** Genetically modified crops can also be engineered to resist diseases caused by viruses, fungi, and bacteria. For example, papaya in Hawaii has been genetically engineered to resist the papaya ringspot virus, preventing crop loss and saving the industry.

Environmental Sustainability

GM crops can play an important role in making agriculture more sustainable and environmentally friendly:

- ☆ **Reduced Pesticide Use:** Crops like Bt cotton and Bt corn, which are resistant to insect pests, require fewer chemical insecticides. This leads to lower chemical inputs, reducing the environmental impact of pesticide use, improving water quality, and benefiting non-target species, including beneficial insects like bees.
- ☆ **Herbicide-Tolerant Crops:** Crops like Roundup Ready soybeans, corn, and cotton are engineered to tolerate herbicides like glyphosate. This allows farmers to control weeds more efficiently, reducing the need for tillage and allowing for conservation tillage practices that help preserve soil health and reduce erosion.
- ☆ **Reduced Greenhouse Gas Emissions:** By reducing the need for chemical inputs like fertilizers and pesticides, GM crops contribute to lower greenhouse gas emissions from agricultural practices. Additionally, conservation tillage and reduced reliance on heavy machinery also result in less fuel consumption and lower carbon footprints.

Enhanced Nutritional Content

GM crops can be engineered to address nutritional deficiencies, especially in developing countries where staple crops like rice, corn, and wheat are central to the diet.

- **Golden Rice:** One of the most famous examples is Golden Rice, which has been genetically modified to produce higher levels of beta-carotene, a precursor to vitamin A. Vitamin A deficiency is a significant public health issue in many developing countries, leading to blindness and other health problems, particularly among children and pregnant women. Golden Rice has the potential to reduce these deficiencies in regions where rice is a primary food source.
- **Biofortified Crops:** Genetic engineering can also increase the levels of essential nutrients such as vitamins, minerals, and amino acids in crops. For example, biofortified maize has been developed with higher levels of lysine and tryptophan, two essential amino acids that are often lacking in the diets of people who rely heavily on maize as a staple food.

Economic Benefits for Farmers

The adoption of genetically modified crops has led to significant economic benefits for farmers, particularly in regions where the crops have been well-received:

- **Increased Yields and Profitability**: The increased resistance to pests, diseases, and environmental stresses has resulted in higher crop yields for farmers. For example, Bt cotton has led to a significant reduction in pest-related losses and an increase in cotton yields, improving farmers' incomes.
- **Reduced Costs for Inputs:** GM crops can reduce the need for expensive chemical inputs, such as pesticides, herbicides, and fertilizers. For instance, herbicide-tolerant crops allow farmers to use a single herbicide, reducing the cost and complexity of weed management.
- **Better Market Access:** Farmers who adopt GM crops may also gain access to better market opportunities, as many large food companies prefer sourcing crops that are disease-resistant or have improved shelf-life or nutritional profiles.

Controversies Surrounding Genetically Modified Crops

Despite the numerous benefits associated with GM crops, there are several concerns and controversies that have sparked public debate. These include issues related to health, the environment, ethics, and social justice.

Environmental Risks

- **Gene Flow and Cross-Breeding:** One of the major concerns regarding GM crops is the potential for gene flow, where genetically modified traits could spread to wild relatives or non-GM crops. This could lead to the development of "superweeds" that are resistant to herbicides or pests, posing a risk to biodiversity and the long-term stability of ecosystems. For example, herbicide-tolerant crops could cross-breed with wild relatives, leading to the development of herbicide-resistant weeds.

- ☆ **Impact on Non-Target Species:** There is concern that transgenic crops, particularly those expressing insecticidal proteins like Bt, could have unintended consequences on non-target species, including beneficial insects, soil organisms, or even wildlife. While studies have shown that Bt crops are relatively safe for non-pest species, there is ongoing research to fully understand their ecological impact.
- ☆ **Loss of Biodiversity:** The widespread adoption of genetically modified crops could result in a reduction in genetic diversity within crops. When a single genetically modified variety is planted over large areas, it may displace traditional or heirloom varieties, reducing the genetic pool available for future breeding and adaptation.

Health Concerns

- ☆ **Allergic Reactions and Toxicity:** One of the most persistent concerns is the potential for allergic reactions or toxic effects from consuming GM crops. Some critics worry that the insertion of foreign genes could produce new allergens or toxins in the crops. While GM crops are rigorously tested for safety, many remain skeptical about long-term health effects and the adequacy of testing.
- ☆ **Unknown Long-Term Effects:** Some opponents argue that the long-term health effects of consuming GM foods are not fully understood, especially with the introduction of newer gene-editing techniques like CRISPR/Cas9. They contend that there could be unintended side effects from altering the genetic makeup of plants in ways that nature would not typically allow, potentially causing changes in the plant's composition that are not immediately obvious.

Socioeconomic and Ethical Concerns

- ☆ **Control by Multinational Corporations:** Another contentious issue is the control of GM crops by large multinational biotech companies, such as Monsanto (now part of Bayer) and Syngenta. Critics argue that these companies have too much control over global food systems, especially in developing countries, and that the high cost of GM seeds and associated technologies could drive small farmers into debt and dependence on a few dominant corporations.
- ☆ **Ethical Issues:** The ethics of genetic modification in agriculture have been debated for decades. Opponents argue that it is unnatural to tamper

with the genetic makeup of organisms, particularly when it comes to food. Some people believe that GM crops pose a moral issue, as they involve altering the genetic structure of living organisms, which can have unintended consequences for ecosystems and human health.

☆ **Impact on Small-Scale Farmers:** While GM crops may provide benefits in terms of increased productivity, critics argue that the widespread use of GM crops could harm small-scale farmers, particularly those in developing countries. Seed patents and licensing agreements often prevent farmers from saving and replanting seeds from GM crops, forcing them to purchase new seeds each season. This could increase the financial burden on farmers, particularly in poorer regions.

Regulation and Oversight

In response to these concerns, GM crops are subject to extensive regulation and oversight by government agencies around the world. In many countries, including the United States and the European Union, GM crops undergo thorough safety assessments to evaluate their environmental impact, potential health risks, and effects on biodiversity before being approved for commercial cultivation. However, the regulatory frameworks vary widely across regions, and many countries, especially in the European Union, have strict rules that limit the cultivation of GM crops.

Plant Tissue Culture and Cloning

Plant tissue culture is a collection of techniques used to grow plant cells, tissues, or organs in a controlled, sterile environment on a nutrient-rich medium. It involves the culture and propagation of plants under in vitro (literally "in glass") conditions, using small pieces of plant material, such as leaves, stems, or even single cells, to generate whole plants. This process is a key tool in plant biotechnology, providing methods for mass propagation, genetic modification, and conservation of rare or endangered plant species. Plant tissue culture and cloning have revolutionized the way we propagate, conserve, and improve plants. By allowing for rapid, mass propagation of disease-free, genetically uniform plants, tissue culture provides a powerful tool for agriculture, horticulture, and conservation. While the process has its challenges, the benefits–ranging from large-scale plant production to the conservation of endangered species–have made tissue. Plant cloning, a significant application of tissue culture, refers to the process of creating genetically identical plants (clones) from a single parent plant. The goal is to replicate a plant's genetic makeup, producing offspring that have the same genetic characteristics, disease resistance, or other beneficial traits as the original plant. Cloning via tissue culture has become increasingly popular in both research and agriculture due to its ability to rapidly propagate plants and preserve desirable traits in high-value crops.

Concepts in Plant Tissue Culture

- **Sterile Environment:** One of the most important aspects of plant tissue culture is maintaining a sterile environment. The introduction of microorganisms like bacteria or fungi can cause contamination, which would inhibit the growth of the plant tissue. To prevent this, the plant tissue is typically sterilized before culture, and the entire process is carried out in sterile conditions, often in laminar flow hoods or controlled environment chambers.
- **Nutrient Medium:** In tissue culture, plant tissues are grown on a nutrient medium that provides the essential nutrients (such as sugars, salts, vitamins, and amino acids) required for growth and development. The medium is often enriched with plant growth regulators (PGRs)–a group of chemical compounds that control cell division, differentiation, and organogenesis. These regulators help guide the development of the plant tissue into shoots, roots, or embryos, depending on the goals of the culture.

Common media used in plant tissue culture include:

- **Murashige and Skoog (MS) medium:** A widely used medium for plant cell and tissue culture.
- **Gamborg's B5 medium:** Another nutrient medium designed for plant cell growth, particularly for suspension cultures.
- **Solid vs. Liquid Media:** Media can be prepared as a gel (using agar) or as a liquid, depending on the desired growth system.

Explants and Initiation

Plant tissue culture typically begins with the use of an explants–a small piece of plant material from a mature plant that is used to start the culture. Explants can be taken from various parts of the plant, including:

- Leaves
- Stems
- Roots
- Meristems (growing tips of the plant)

The explant is placed on the nutrient medium in a sterile container, and under the right conditions, it will begin to divide and grow. Explant types can influence the success of tissue culture; for example, meristematic tissues (which are actively dividing cells) are often used for cloning because they have high regenerative capacity.

Stages of Plant Tissue Culture

The process of plant tissue culture generally involves several stages:

- ☆ **Callus Formation:** In the first stage, explants are placed in a nutrient medium containing appropriate plant growth regulators. The explants will start to divide and form a mass of undifferentiated cells called a callus. Callus formation is the initial step in tissue culture and is essential for the subsequent development of the plant.
- ☆ **Shoot and Root Induction**: Once a callus has formed, it is transferred to a medium with different growth regulators to promote the formation of shoots (vegetative organs) and roots. The medium used for shoot induction typically contains cytokinins (which promote cell division) and a lower concentration of auxins (which encourage root formation). Conversely, for root induction, the medium typically contains a higher concentration of auxins and lower cytokinin levels.
- ☆ **Plant Regeneration:** In the final stage, the callus undergoes differentiation to form fully functional shoots and roots. The plantlets are then transferred to a soil-like medium for acclimatization and eventually moved to soil or another growing environment for full development into mature plants.

Applications of Plant Tissue Culture

- ☆ **Cloning and Propagation:** The most common application of tissue culture is cloning–the production of genetically identical plants. By using tissue culture techniques, a single plant can be cloned multiple times, creating large numbers of plants that are genetically uniform. This is particularly useful in agriculture, horticulture, and forestry for propagating plants that are difficult to grow from seeds, have long growing cycles, or are prone to diseases.

  Examples of plants commonly propagated through tissue culture include:

  - ❖ **Ornamental plants** like roses, orchids, and lilies
  - ❖ **Fruit crops** such as bananas, apples, and strawberries
  - ❖ **Crops with desirable traits** like high yield or disease resistance
- ☆ **Genetic Improvement and Modification:** Tissue culture is also used as a foundation for the genetic modification of plants. Through genetic engineering, genes can be introduced into plant cells via tissue culture, creating genetically modified (GM) plants. Once the gene has been incorporated, tissue culture methods help regenerate complete plants that express the desired trait. For example, tissue culture is used in conjunction with techniques like Agrobacterium-mediated transformation or gene gun technology to create transgenic plants with traits such as resistance to pests (*e.g.*, Bt crops), tolerance to herbicides (*e.g.*, Roundup Ready crops),

or enhanced nutritional content (*e.g.*, Golden Rice).

- **Conservation of Endangered Species:** Tissue culture plays a critical role in the conservation of endangered plant species. By using tissue culture techniques like somatic embryogenesis and cryopreservation, plant biotechnologists can store the genetic material of rare or endangered plants, ensuring their survival even if the plant population is at risk of extinction in the wild. Tissue culture also allows the reproduction of plants from very small samples, which is particularly important for species that cannot be easily propagated by traditional methods.
- **Disease-Free Plant Production:** One of the key benefits of tissue culture is its ability to produce disease-free plants. Plant tissues can be sterilized before culturing, and during the regeneration phase, infected plants or plant parts can be avoided or eliminated. This is especially important for crops like potatoes, bananas, and grapes, which are often susceptible to viral or bacterial diseases that can spread through infected seeds or vegetative propagules. By starting with disease-free tissue, plant breeders can ensure healthier and more productive crops.
- **Hybridization and Somatic Hybridization;** Tissue culture is also employed in somatic hybridization, a technique in which two different plant species are fused at the cellular level to create a hybrid plant with characteristics from both parent plants. This can be done using protoplast fusion, where the cell walls of two plant cells are removed, and the resulting protoplasts are fused together to form a hybrid. Somatic hybridization can be used to combine traits from different species that would not normally crossbreed through sexual reproduction, such as creating disease-resistant crops or plants with enhanced yields.

Advantages of Plant Tissue Culture and Cloning

- **Mass Propagation:** Tissue culture allows for the rapid production of large numbers of plants from a small amount of initial plant material. This is especially valuable for the commercial production of plants that are difficult to propagate through traditional means.
- **Genetic Uniformity:** Cloning ensures that all plants are genetically identical, which is essential for maintaining uniformity in crop traits such as yield, disease resistance, or flower color in ornamental plants.
- **Disease-Free Plants:** Tissue culture techniques are highly effective in producing disease-free plants by eliminating pathogens that may be present in seeds, cuttings, or tubers.
- **Conservation of Genetic Resources:** Tissue culture enables the preservation of plant genetic material, particularly for endangered species, rare cultivars, or important crop varieties that may be at risk of extinction.

☆ **Overcoming Seasonal Limitations:** Since tissue culture occurs in a controlled environment, it can be done year-round, independent of external growing conditions like seasonality or climate.

☆ **Germplasm Storage:** Through cryopreservation, plant genetic material can be stored indefinitely, providing a valuable resource for future breeding and conservation efforts.

Challenges and Limitations

Despite its advantages, plant tissue culture and cloning face several challenges:

☆ **Cost:** Setting up and maintaining tissue culture labs can be expensive due to the need for specialized equipment, sterile environments, and skilled labor.

☆ **Somaclonal Variation:** Although cloning creates genetically identical plants, sometimes mutations occur during the tissue culture process, leading to genetic variation in the plants (called somaclonal variation). This can lead to undesirable traits in the offspring.

☆ **Acclimatization:** Plants grown in tissue culture need to be carefully acclimatized to external environments (*e.g.*, soil, climate), as they are initially grown in controlled lab conditions. Improper acclimatization can lead to poor survival rates when transplanted into the field.

☆ **Contamination:** Maintaining a sterile culture is essential, but even minor contamination by bacteria or fungi can ruin the culture and require restarting the entire process.

The Future of Plant Biotechnology

Plant biotechnology is evolving rapidly, with the potential to revolutionize agriculture, food production, and environmental sustainability on a global scale. As the world faces significant challenges such as climate change, population growth, dwindling natural resources, and rising food insecurity, the future of plant biotechnology holds great promise in addressing these pressing issues. With advances in genetic engineering, gene editing technologies like CRISPR, synthetic biology, and improved plant tissue culture techniques, the scope of what is possible in plant biotechnology is expanding dramatically. This new era of innovation offers the possibility of designing crops that can thrive in harsh environments, provide more nutritious food, and reduce the environmental impact of farming. However, these advancements also raise important questions about ethics, regulation, and long-term ecological effects, making the future of plant biotechnology a topic of intense research and debate. The future of plant biotechnology is filled with immense potential to address some of the most urgent challenges facing humanity. From creating crops that are more resilient to climate change and more nutritious to developing agricultural practices that are more sustainable and environmentally friendly, plant biotechnology holds the promise of transforming the way we grow food and manage natural resources. However, for these advancements to be fully realized, careful attention must be paid to the ethical, regulatory, and social implications of these technologies. With the right balance of innovation and responsibility, plant biotechnology could play a crucial role in feeding the world's growing population while ensuring a healthier and more sustainable planet for future generations.

- ☆ **Climate-Resilient Crops for a Changing World:** One of the most exciting prospects for the future of plant biotechnology is the development of climate-resilient crops. As global temperatures rise and weather patterns become more unpredictable, agriculture is increasingly vulnerable to environmental stressors like drought, heat, flooding, and salinity. Traditional crop varieties often struggle to adapt to these challenges, leading to reduced yields and increased pressure on food production. Biotechnology, however, has the potential to provide solutions that could help crops survive and even thrive in extreme conditions. Through genetic engineering, scientists are working to develop crops with enhanced drought tolerance, heat resistance, and flood tolerance by introducing or modifying genes that enable better water retention, stress resistance, and root development. For example, genetic modifications in crops like corn, rice, and wheat are already underway to enhance their ability to withstand heat and drought stress, which could be crucial for food security in the face of climate change. Salinity-tolerant rice has been developed in experimental trials, enabling rice to grow in areas with higher soil salinity, which would otherwise be unsuitable for cultivation. The future of plant biotechnology will likely focus on further refining these traits, allowing crops to be grown in areas that are currently considered marginal for agriculture, such as arid deserts, saline coastal lands, and regions experiencing the effects of flooding and other extreme weather events.
- ☆ **Improved Nutritional Content and Biofortification:** As the global population continues to grow, there is an increasing need for not only more food but also food that is more nutritionally dense to combat malnutrition, especially in developing countries. The future of plant biotechnology holds exciting potential for biofortification, which is the process of enhancing the nutritional content of crops through genetic modification or breeding techniques. This is particularly crucial in addressing micronutrient deficiencies, which affect billions of people worldwide. In the future, crops can be engineered to contain higher levels of vitamins, minerals, and essential amino acids. For instance, Golden Rice, which is genetically modified to produce higher levels of beta-carotene (a precursor to Vitamin A), has already been developed as a means to combat Vitamin A deficiency, a major cause of blindness and other health issues, particularly in children in parts of Asia and Africa. Similarly, iron-

fortified beans, zinc-enriched wheat, and protein-enhanced maize are being developed to address deficiencies in iron, zinc, and protein, which are among the most common nutritional deficiencies globally. In the future, plant biotechnology could lead to even more advanced biofortified crops that address multiple deficiencies at once, potentially creating "superfoods" that can help eliminate malnutrition in regions where food diversity is limited. Furthermore, the use of CRISPR-based gene editing technologies could allow for more precise and efficient enhancement of these nutrients without introducing foreign DNA, which could appeal to both consumers and regulatory agencies concerned about the safety of transgenic crops.

- ☆ **Sustainability and Reducing Agriculture's Environmental Footprint:** Agriculture, while essential for food production, is a major contributor to environmental degradation. Greenhouse gas emissions, soil erosion, water overuse, and biodiversity loss are some of the environmental challenges associated with conventional farming practices. Plant biotechnology offers the potential to reduce the environmental footprint of agriculture in several ways. Biotech crops can be engineered to require fewer inputs, such as water, fertilizers, and pesticides, which in turn reduces pollution, minimizes soil degradation, and conserves water resources. For example, crops engineered to be resistant to pests, like Bt cotton and Bt corn, have led to significant reductions in pesticide use, which not only lowers the environmental impact of farming but also reduces costs for farmers. Additionally, genetically modified crops with herbicide tolerance, such as glyphosate-tolerant soybeans, enable farmers to use herbicides more effectively, reducing the need for tillage and promoting conservation tillage practices that improve soil health and reduce erosion. In the future, plant biotechnology will likely play an increasingly critical role in creating crops that are not only more productive but also more sustainable. Nitrogen-efficient crops, for instance, could help reduce the need for synthetic fertilizers, which are a major source of greenhouse gas emissions and water pollution. Similarly, crops that require less water or are more efficient in their water use could be vital in areas facing water scarcity, thus conserving precious water resources.
- ☆ **Gene Editing and Precision Agriculture:** The future of plant biotechnology will also be shaped by the rise of gene editing technologies, particularly CRISPR/Cas9 and other genome-editing techniques. These tools allow scientists to make very precise modifications to the plant genome, enabling the development of crops with specific traits without introducing foreign genes. This is a major step forward compared to traditional genetic modification, which typically involves the insertion of genes from different species. With gene editing, scientists can target specific genes responsible for desirable traits, such as pest resistance,

drought tolerance, or enhanced nutritional content, and modify them directly.Gene editing also holds promise for precision agriculture, where plants can be customized to meet the exact needs of local environments, increasing productivity and minimizing waste. For example, gene-edited plants could be tailored to grow more efficiently in different soil types, withstand specific pests and diseases, or mature more quickly, all of which would contribute to higher yields with fewer inputs.One of the most exciting potential applications of gene editing in agriculture is the development of crops that can be used as biofuels, offering an alternative to fossil fuels. Researchers are exploring gene-editing techniques to improve the yield and efficiency of crops like switchgrass, algae, and oilseeds that could be used for renewable energy production. This could help reduce global reliance on oil, mitigate climate change, and provide new economic opportunities for farmers.

- **Ethical, Regulatory, and Public Perception Challenges:** Despite the immense potential of plant biotechnology, there are significant challenges related to the ethics, regulation, and public perception of genetically modified crops. As the technology advances, the future of plant biotechnology will be shaped by the regulatory frameworks that govern its development and use. Different countries have different approaches to the regulation of GM crops, and the level of public acceptance varies widely. In some countries, genetically modified organisms (GMOs) face stringent regulatory hurdles and consumer skepticism, while in others, GM crops are embraced as a solution to food security challenges. The ethical implications of genetic modification, particularly with the use of gene editing technologies like CRISPR, are also the subject of ongoing debate. Questions about the potential unintended consequences of altering plant genomes, especially for crops that will be consumed by humans, raise concerns about long-term safety and ecological impact. Furthermore, the use of biotechnology in agriculture can exacerbate concerns about corporate control over seed patents and the monopolization of the food supply by a few multinational corporations. The future of plant biotechnology will depend not only on scientific breakthroughs but also on how well the public, regulators, and policymakers navigate these ethical and societal challenges. It will be crucial to ensure that advances in plant biotechnology are accessible to all, particularly small-scale farmers and developing nations, and that they do not disproportionately benefit wealthy countries or corporations at the expense of global equity and sustainability.

Chapter 11

Evolutionary and Ecological Perspectives on Plant Physiology

Introduction

Plant physiology, when examined through the lenses of evolutionary and ecological perspectives, unveils a profound understanding of how plants function, adapt, and interact with their environments. Plants, as primary producers, are central to the Earth's ecosystems, providing the foundation for most food chains and playing key roles in regulating atmospheric gases, nutrient cycling, and climate stability. Evolutionarily, plant physiology is shaped by millions of years of natural selection, during which plants have developed a remarkable array of adaptive strategies to survive and thrive in a diverse range of habitats. These adaptations include specialized mechanisms for water and nutrient uptake, photosynthesis efficiency, reproduction, and responses to environmental stressors like temperature extremes, drought, and herbivory. Ecologically, plant physiology is closely intertwined with interactions between plants and other organisms, such as herbivores, pollinators, and symbiotic microorganisms, as well as abiotic factors like soil composition, light, and water availability. By integrating evolutionary and ecological insights, we can better understand how plants have evolved to meet the challenges of their environments and how they continue to influence, and be influenced by, the broader ecological systems in which they exist. This holistic perspective not only illuminates the intricate biological mechanisms behind plant function but also underscores the dynamic relationship between plants and the changing global environment.

Plant Evolution and Adaptation to Diverse Environments

Plant evolution and adaptation to diverse environments is a fascinating story of survival, innovation, and diversification. Over millions of years, plants have evolved an extraordinary range of traits and strategies that allow them to thrive in a vast array of habitats, from the most arid deserts to lush tropical rainforests, and from freezing polar regions to high-altitude mountain slopes. These adaptations have been driven by natural selection in response to varying environmental pressures, including light, water, temperature, nutrient availability, and interactions with other organisms.

- **Evolution of Vascular Systems:** One of the key evolutionary steps in plant history was the development of vascular tissue–xylem and phloem–which enabled plants to move water and nutrients efficiently over long distances. The emergence of vascular plants over 400 million years ago was a critical adaptation that allowed plants to colonize land and grow to larger sizes. This was particularly important as plants had to contend with the challenge of water retention and the need to transport nutrients across increasingly diverse terrestrial environments.
- **Water Conservation Mechanisms:** One of the most significant adaptive strategies in plants is the evolution of mechanisms to minimize water loss. In arid environments, where water is scarce, plants have evolved specialized features such as waxy cuticles, stomatal regulation, and deep root systems. For instance, xerophytes, like cacti, have thickened cuticles and specialized structures called CAM (Crassulacean Acid Metabolism) to store water and photosynthesize at night, reducing water loss during the heat of the day. Conversely, hydrophytes (aquatic plants) have adaptations such as air sacs to float, enabling them to thrive in waterlogged conditions where oxygen might be limited.
- **Temperature Tolerance:** Plants have evolved strategies to cope with temperature extremes, which are crucial for survival in environments ranging from scorching deserts to freezing tundras. Thermophilic plants, such as those found in volcanic regions, have adapted to high temperatures by stabilizing their cellular enzymes. In contrast, cold-hardy plants or cryophytes, such as alpine species, can survive sub-zero temperatures through antifreeze proteins, changes in cell membrane composition, and the ability to enter dormancy during extreme cold.
- **Photosynthetic Adaptations:** Plants in different environments have also evolved various types of photosynthesis to maximize energy capture while minimizing water loss. The most common form, C3 photosynthesis, is efficient under cool, moist conditions. However, in hot, dry environments, plants have evolved C4 and CAM photosynthesis. C4 plants, like maize and

sugarcane, use a different biochemical pathway to concentrate carbon dioxide, making them more efficient in hot climates. CAM plants, like succulents, open their stomata at night to take in carbon dioxide, reducing water loss during the heat of the day.

- **Reproductive Adaptations:** Plants have evolved a wide variety of reproductive strategies to ensure their survival and propagation across diverse environments. In areas where pollinators are scarce, wind-pollinated plants, such as grasses and many conifers, produce large amounts of pollen that can be carried by the wind. In contrast, in ecosystems with abundant pollinators, plants have developed intricate floral structures and scents to attract specific pollinators. Additionally, plants have evolved diverse seed dispersal mechanisms, such as using wind, water, or animals, to ensure that their offspring can reach new habitats where conditions might be more favorable.
- **Ecological Interactions and Coevolution:** Plants do not evolve in isolation but are in constant interaction with other organisms. Over time, plants have coevolved with herbivores, pollinators, and symbiotic microbes. For example, many plants have developed chemical defenses–such as toxins or bitter compounds–to deter herbivores. Similarly, plants like legumes have evolved symbiotic relationships with nitrogen-fixing bacteria, which help them obtain nutrients in nitrogen-poor soils.
- **Evolution of Plant Diversity:** Plant evolution has led to the remarkable diversity of plant forms and functions that we see today. This diversity reflects both the different ecological niches plants occupy and the various adaptive strategies they have developed. From tiny mosses that thrive in shaded, moist environments to towering trees that dominate forest canopies, plants have adapted to virtually every terrestrial and aquatic environment on Earth. The evolution of flowers, fruits, and seeds has further facilitated plant diversification and reproduction, enabling plants to exploit a wide range of habitats and ecological niches.

Plant evolution and adaptation is a dynamic and ongoing process, shaped by a complex interplay of genetic changes, environmental factors, and ecological interactions. The diversity of plant forms and the multitude of strategies plants use to survive in diverse environments are a testament to their remarkable evolutionary success and their essential role in maintaining Earth's ecosystems.

Ecophysiology: The Role of Physiology in Ecology

Ecophysiology is the study of how an organism's physiological processes are shaped by and, in turn, influence its interactions with the environment. This interdisciplinary field bridges the gap between physiology (how organisms function) and ecology (how organisms interact with their environment). Understanding ecophysiology is crucial for explaining how organisms survive, reproduce, and evolve in diverse and changing ecosystems. Ecophysiology is a crucial discipline for understanding the survival strategies of organisms in their environment. It combines insights from biology, ecology, and environmental science to predict how species will cope with challenges in their habitats. In the face of ongoing environmental change, ecophysiology offers valuable tools for managing biodiversity and conserving ecosystems. By understanding the physiological underpinnings of ecological interactions, we can better predict, protect, and manage the living world around us.

Concepts in Ecophysiology

Physiological Adaptations to the Environment

Organisms face various environmental challenges such as temperature extremes, water availability, salinity, oxygen levels, and the availability of food. Ecophysiology explores how organisms have evolved specific physiological mechanisms to cope with these stressors. Examples include:

- ☆ **Thermoregulation:** The ability to maintain body temperature, either through behavioral adaptations (*e.g.*, seeking shade) or physiological processes (*e.g.*, sweating or shivering).
- ☆ **Water Balance:** Adaptations to conserve or manage water, such as the production of concentrated urine in desert animals or the ability of plants to close stomata to prevent water loss.
- ☆ **Oxygen Uptake:** In aquatic systems, fish and other organisms have evolved ways to extract oxygen from water, while high-altitude animals have specialized circulatory and respiratory systems to cope with low oxygen availability.

Energy Acquisition and Allocation

Energy acquisition through processes like photosynthesis in plants or foraging in animals is another critical area of ecophysiology. How organisms allocate this energy to growth, reproduction, maintenance, and response to environmental challenges influences their fitness. For example:

- ☆ **Photosynthesis and Light Capture:** In plants, the efficiency of photosynthesis depends on environmental factors such as light intensity, temperature, and carbon dioxide levels.
- ☆ **Foraging Strategies:** Animals have evolved different foraging strategies based on metabolic demands, environmental constraints, and food availability.

Stress Physiology

Organisms often encounter environmental stresses (*e.g.*, drought, temperature extremes, predation) that trigger physiological responses. Understanding stress responses, such as the production of stress hormones (*e.g.*, cortisol in animals), helps explain how organisms cope with changing conditions. Some key processes include:

- **Acclimatization:** Short-term, reversible physiological adjustments to environmental changes, such as increased production of heat-shock proteins in response to high temperatures.
- **Adaptation:** Long-term evolutionary changes in an organism's physiology that enhance survival in specific environments. For example, the evolution of salt-excreting glands in marine birds.

Interactions Between Organisms and their Environment

Ecophysiology also examines how environmental factors (*e.g.*, climate, soil type, water chemistry) influence ecological interactions such as competition, predation, and mutualism. For example:

- **Plant-Pollinator Interactions:** Physiological changes in flowering time or nectar production in response to temperature or light can affect plant-pollinator dynamics.
- **Predator-Prey Dynamics:** The energy expenditure of predators, their hunting strategies, and the physiological adaptations of prey (*e.g.*, camouflaging, flight speed) are key in understanding ecological dynamics.

Global Change and Ecophysiology

With rapid changes in climate, habitat destruction, and pollution, ecophysiology plays an important role in understanding how species will respond to these shifts. For example:

- **Climate Change:** Organisms may experience altered temperature regimes, altered precipitation patterns, and changing food availability. Physiological research can predict how species might adapt, acclimate, or face extinction under changing climatic conditions.
- **Pollution Tolerance:** Some species have developed physiological mechanisms to tolerate pollutants, such as heavy metals or acid rain, while others may be more vulnerable.

Ecophysiology in Action: Examples

- **Desert Adaptations:** Many desert animals and plants have evolved unique physiological adaptations to survive extreme conditions. For example, the kangaroo rat has highly efficient kidneys that conserve water, and certain cactus species can store large amounts of water in their tissues, enabling them to survive prolonged dry periods.
- **Aquatic Adaptations:** Fish living in cold, oxygen-poor waters may have specialized hemoglobin that can efficiently bind oxygen, whereas

fish in warmer waters may have higher metabolic rates and different enzyme activities. Marine mammals, like seals, have adaptations in their circulatory systems to conserve oxygen during deep dives.

- **High-altitude Species:** Animals and plants that live at high altitudes must cope with lower oxygen levels and extreme weather. For instance, the Tibetan antelope has evolved larger lungs and a higher concentration of hemoglobin to efficiently extract oxygen from the thin air at high elevations.

Ecophysiology and Conservation

Understanding the physiological requirements and limitations of organisms is crucial for conservation efforts. As human-induced environmental changes continue to accelerate, ecophysiology can provide insights into:

- **Predicting Species' Vulnerability to Climate Change:** Species with narrow physiological tolerances are more likely to be affected by rapid climate change. By studying the physiological limits of species, ecophysiologists can help predict their ability to adapt.
- **Restoration Ecology:** When restoring damaged ecosystems, understanding the physiological needs of native species can inform planting or reintroducing efforts to ensure that organisms are able to thrive in the altered environment.

Plant Interactions with Other Organisms: Symbiosis and Competition

Plants do not live in isolation; they are surrounded by and interact with various other organisms that can either help or hinder their growth, survival, and reproduction. These interactions are diverse and complex, involving relationships with other plants, animals, fungi, bacteria, and even other types of microorganisms. In ecology, plant interactions with other organisms can be broadly categorized into symbiosis (mutually beneficial interactions) and competition (antagonistic interactions), both of which play a significant role in shaping plant communities and ecosystems.

Symbiosis in Plants

Symbiosis refers to any close, long-term biological interaction between two different species. The relationship can be beneficial, neutral, or harmful to one or both of the involved species. Symbiotic interactions in plants are often classified into three main types: mutualism, commensalism, and parasitism. In plants, the most important and common symbiotic interactions are mutualism, where both partners benefit from the relationship.

Mutualism

In mutualistic symbiosis, both species benefit from the interaction, and these relationships are often essential for the survival or success of one or both partners. Some key examples of plant mutualisms include:

Mycorrhizal Associations (Plant-Fungi Symbiosis)

One of the most well-known mutualistic relationships in plants is the association between plant roots and mycorrhizal fungi. In this relationship, the fungus colonizes the plant's root system, increasing the surface area for water and nutrient absorption, particularly phosphorus and nitrogen. In exchange, the plant provides the fungus with carbohydrates (sugars), which the fungus cannot produce on its own.

Types of Mycorrhizae

- ✰ *Ectomycorrhizae*: The fungus forms a sheath around the root surface and grows between the root cells.
- ✰ *Arbuscular mycorrhizae (AMF)*: The fungus penetrates the plant root cells and forms structures called arbuscules where nutrient exchange occurs.

This relationship is vital in many ecosystems, especially in nutrient-poor soils, where plants rely heavily on fungi for nutrient uptake.

Rhizobial Symbiosis (Nitrogen-Fixing Bacteria and Legumes)

Certain plants, particularly legumes (*e.g.*, peas, beans, and clover), form a mutualistic relationship with nitrogen-fixing bacteria in the genus *Rhizobium*. These bacteria live in nodules on the plant's roots and convert atmospheric nitrogen (N) into ammonia (NH_3), a form of nitrogen that plants can utilize for growth. In return, the plant supplies the bacteria with carbohydrates and other nutrients.

This relationship significantly enhances soil fertility by increasing the availability of nitrogen, benefiting both the host plants and the surrounding ecosystem.

Pollination (Plant-Insect Symbiosis)

Pollination is a form of mutualism between plants and pollinators such as bees, butterflies, birds, and bats. The plant provides nectar (a food resource) and pollen (a protein-rich food for many pollinators), while the pollinator helps the plant reproduce by transferring pollen from one flower to another, facilitating fertilization.

This interaction is crucial for the reproduction of many plants and the maintenance of biodiversity, as it enables genetic diversity through cross-pollination.

Seed Dispersal (Plants and Animals)

Many plants rely on animals for seed dispersal, a mutualistic relationship where plants benefit from the movement of seeds away from the parent plant, reducing competition for resources. Animals, in return, often consume fruits or seeds, providing the plant with an effective means of dispersing its offspring. For example, birds eat berries and excrete seeds in new locations, while squirrels may cache seeds that later grow into new plants.

Commensalism

In commensalism, one species benefits from the interaction, while the other is neither helped nor harmed. Examples in plant interactions include:

Epiphytic Plants (Epiphytes)

Epiphytes, such as many species of orchids, mosses, and bromeliads, grow on the surfaces of trees in rainforests. They derive physical support from the tree but do not take nutrients from it. Epiphytes benefit from the elevated position, gaining better access to sunlight and moisture. The tree is largely unaffected by the presence of the epiphyte.

Parasitism

Parasitism is a form of symbiosis where one species benefits at the expense of the other. Parasitic plants absorb nutrients from their host plants, often harming them in the process. Examples of parasitic plants include:

- **Mistletoe:** Mistletoe is a well-known parasitic plant that attaches to the branches of trees and shrubs. It taps into the host's vascular system to absorb water, nutrients, and sugars, often weakening the host plant over time.
- **Dodder (Cuscuta):** Dodder is a parasitic vine that lacks chlorophyll and wraps itself around the stems or branches of other plants. It absorbs nutrients from the host plant and can severely damage or kill it if left unchecked.

Competition in Plants

Competition occurs when two or more plants vie for the same limited resources, such as light, water, nutrients, or space. Competition can occur between plants of the same species (intraspecific competition) or between different species (interspecific competition). This interaction can be detrimental to one or both species involved, as it limits growth, reproduction, and survival.

Intraspecific Competition

Intraspecific competition refers to competition between individuals of the same species. This type of competition is often intense because individuals have identical resource requirements, such as light, water, and nutrients. Intraspecific competition can occur when:

- ☆ **Seedlings compete for light and soil nutrients.**
- ☆ **Roots and shoots of neighboring plants invade each other's space.**

This competition may result in self-thinning, where only the strongest individuals survive and grow, leading to a more densely packed population of plants in a particular area.

Interspecific Competition

Interspecific competition involves competition between individuals of different species. This can occur in various ways:

- ☆ **Aboveground Competition for Light:** Plants compete for light in a crowded ecosystem. Tall, fast-growing plants, such as trees, often overshadow smaller, shade-intolerant plants, limiting their access to sunlight.
- ☆ **Belowground Competition for Water and Nutrients:** Roots of neighboring plants may extend into the same soil, competing for water, nitrogen, phosphorus, and other nutrients. Some plants, like grasses, may have shallow root systems, while others, like trees, have deep roots to access different water or nutrient layers in the soil.

One example of interspecific competition is the interaction between *Cucurbita* species (squash plants) and *Zea mays* (corn), where both species may compete for light and soil nutrients in agricultural systems.

Allelopathy (Chemical Competition)

Some plants compete indirectly by releasing chemicals into the environment that inhibit the growth of nearby plants. This phenomenon is known as allelopathy. Certain plants, such as black walnut trees (*Juglans nigra*), release juglone, a toxic compound, from their roots, leaves, and husks. This chemical can inhibit seed germination and growth of other plants in the area, giving the walnut tree a competitive advantage.

Plant Interactions and Community Structure

The nature of plant interactions–whether symbiotic or competitive–can have profound impacts on the structure and diversity of plant communities. Symbiotic relationships, especially mutualisms, promote the success and persistence of plants, facilitating species coexistence and ecological balance. On the other hand, competitive interactions can limit species diversity, as dominant species often outcompete others for resources, leading to competitive exclusion.

- ☆ **Facilitation:** Some plants can positively affect the growth of other species through their interactions. For example, nurse plants may provide shade and protection from harsh environmental conditions, allowing other species to establish in otherwise inhospitable environments.

- ☆ **Competitive Exclusion Principle:** This principle suggests that two species that compete for identical resources cannot coexist indefinitely in the same ecological niche. One species will outcompete the other, leading to the exclusion of the less competitive species.

The interactions between plants and other organisms–whether mutualistic, competitive, or parasitic–are integral to understanding plant ecology. These relationships shape the structure of ecosystems, influence the biodiversity of plant communities, and can determine the success or failure of individual species in a given environment. Through mutualism, plants can access resources, enhance reproduction, and foster cooperation with other species. Conversely, through competition, plants must navigate the challenges of limited resources, sometimes leading to the exclusion of less competitive species. Together, symbiosis and competition help define the dynamics of natural and human-modified ecosystems.

Ecological Adaptations: CAM and C4 Photosynthesis

Plants have evolved a variety of strategies to optimize their ability to capture and use carbon dioxide (CO_2) for photosynthesis, the process by which they convert light energy into chemical energy. These strategies are particularly important in environments where water is scarce, temperatures are high, or light intensity fluctuates, making photosynthesis more challenging. CAM (Crassulacean Acid Metabolism) and C4 photosynthesis are two key ecological adaptations that have evolved in plants to cope with environmental stresses such as water limitation and high temperature. Both adaptations modify the typical photosynthetic pathway to help plants efficiently capture CO_2 and minimize water loss, but they do so in different ways.

Overview of Photosynthesis

In most plants, photosynthesis occurs in two main stages:

1. **Light-dependent reactions:** These reactions take place in the thylakoid membranes of the chloroplasts, where light energy is used to produce ATP and NADPH.
2. **Calvin Cycle (Light-independent reactions):** In the stroma of the chloroplasts, ATP and NADPH from the light reactions are used to fix CO_2 into organic compounds, primarily glucose.

In C3 photosynthesis (the most common form), CO_2 enters the plant via the stomata and is fixed into a 3-carbon molecule by the enzyme RuBisCO. However, under conditions of high light intensity, high temperature, and low water availability, $RuBisCO_2$ can also react with oxygen (O_2) instead of CO_2, leading to a wasteful process called photorespiration. This reduces the efficiency of photosynthesis, especially under stressful conditions.

CAM (Crassulacean Acid Metabolism) Photosynthesis

CAM photosynthesis is an adaptation to arid and semi-arid environments, where water is limited. CAM plants, such as cacti, succulents, and agave, employ a unique strategy to reduce water loss while still capturing CO_2 for photosynthesis.

Features of CAM Photosynthesis

Temporal Separation of CO_2 Fixation and Photosynthesis: CAM plants separate the processes of CO_2 uptake and the Calvin Cycle by time, rather than space (as seen in C4 plants).

- ✰ **Nighttime (Stomata Open):** CAM plants open their stomata at night when the temperature is lower, and humidity is higher, reducing water loss. During the night, the plant absorbs CO_2 from the air and stores it as **malic acid** (or other organic acids) in vacuoles.
- ✰ **Daytime (Stomata Closed):** During the day, the stomata close to conserve water. The stored malic acid is decarboxylated to release CO_2, which is then fixed in the Calvin Cycle using the ATP and NADPH generated during the day.

Benefits of CAM Photosynthesis

- ✰ **Water Conservation:** By opening the stomata only at night, CAM plants minimize water loss through transpiration during the hot, dry daytime hours.
- ✰ **Efficient CO_2 Utilization:** Despite stomatal closure during the day, CAM plants can still perform photosynthesis because they have a store of CO_2 in the form of organic acids.

Examples of CAM Plants

- ✰ Cacti (*e.g., Saguaro cactus*)
- ✰ Agave
- ✰ Pineapple
- ✰ Aloe
- ✰ Crassula (*e.g.,* jade plant)

Ecological Role of CAM Photosynthesis

CAM plants are well-suited for desert and drought-prone environments, where water conservation is essential. By performing photosynthesis while limiting water loss, these plants can survive in environments where other types of plants (such as C3 plants) would not be able to thrive. CAM is a particularly important adaptation in plants that are also capable of storing water in their tissues, as seen in succulents.

C4 Photosynthesis

C4 photosynthesis is another adaptation to high-temperature, high-light environments and is common in plants found in tropical or subtropical regions. C4 plants, such as maize (corn), sugarcane, and switchgrass, have evolved a more efficient way to fix CO_2, reducing the effects of photorespiration.

Features of C4 Photosynthesis

Spatial Separation of CO_2 Fixation and Calvin Cycle

Unlike CAM plants, which separate CO_2 fixation and the Calvin Cycle by time, C4 plants separate these processes by cell type. The process occurs in two types of cells:

- ☆ **Mesophyll Cells:** CO_2 is first captured in the mesophyll cells (the outer cells near the leaf surface) and is incorporated into a 4-carbon compound, usually oxaloacetate or malate, through the enzyme PEP carboxylase. This enzyme has a higher affinity for CO_2 and does not bind to oxygen, thus avoiding photorespiration.
- ☆ **Bundle Sheath Cells:** The 4-carbon compound is then transported into the bundle sheath cells, where it is decarboxylated to release CO_2. This CO_2 is then used in the Calvin Cycle, which occurs in the bundle sheath cells, where RuBisCO$_2$ is located. By concentrating CO_2 around RuBisCO, C4 plants reduce the chances of photorespiration and enhance the efficiency of photosynthesis.

Benefits of C4 Photosynthesis

- ☆ **Reduced Photorespiration:** The spatial separation between the mesophyll and bundle sheath cells allows C4 plants to concentrate CO_2 around RuBisCO, minimizing the wasteful process of photorespiration.
- ☆ **High Efficiency in Hot, Sunny Environments:** C4 plants can perform photosynthesis more efficiently at high temperatures and intense sunlight, as they do not lose carbon to photorespiration. This makes them highly productive in warm, sunny climates.
- ☆ **Water Use Efficiency:** C4 plants are generally more water-efficient than C3 plants because they can maintain a high rate of photosynthesis with fewer stomatal openings, thus reducing water loss through transpiration.

Examples of C4 Plants

- ☆ **Maize** (corn)
- ☆ **Sugarcane**
- ☆ **Sorghum**
- ☆ **Switchgrass**
- ☆ **Millets**
- ☆ **Certain grasses** (*e.g.*, Bermuda grass, ryegrass)

Ecological Role of C4 Photosynthesis

C4 plants are well adapted to tropical, subtropical, and savanna ecosystems, where high temperatures and intense sunlight make traditional C3 photosynthesis inefficient due to increased photorespiration. They are often the dominant plants in grasslands and agricultural systems in warm climates, contributing significantly to biomass production.

Table 11.1. Comparison between CAM and C4 Photosynthesis

Feature	*CAM Photosynthesis*	*C4 Photosynthesis*
Type of Environmental Stress	Water scarcity, arid environments	High temperature, high light intensity, low CO_2 availability
Separation Mechanism	Temporal (night and day)	Spatial (mesophyll and bundle sheath cells)
CO_2 Fixation	At night (stomata open)	In mesophyll cells (CO_2 initially fixed into 4-carbon compounds)
Primary Benefit	Water conservation	Reduced photorespiration, increased photosynthetic efficiency
Examples of Plants	Cacti, succulents, pineapple, agave	Corn, sugarcane, sorghum, switchgrass

Ecological and Agricultural Implications

Both CAM and C4 photosynthesis allow plants to perform photosynthesis more efficiently in challenging environments, but each has its specific advantages:

- ☆ CAM plants are ideal for drought-prone regions where water is the limiting factor. Their ability to open stomata only at night minimizes water loss and makes them highly suited to dry conditions.
- ☆ C4 plants are more suited to high-temperature environments where high light and heat levels can lead to photorespiration in C3 plants. They are highly efficient in tropical or subtropical climates and are important in agriculture, especially for crops like maize and sugarcane.

These photosynthetic adaptations are not only crucial for plant survival in specific ecological niches but also have important implications for agriculture, carbon sequestration, and climate change resilience. By understanding how plants adapt to their environments through these specialized forms of photosynthesis, scientists and farmers can improve crop productivity, enhance water use efficiency, and develop more resilient agricultural systems.

Co-evolution of Plants and Pollinators

The relationship between plants and their pollinators is one of the most remarkable examples of co-evolution in nature. Co-evolution refers to the reciprocal evolutionary changes that occur between two or more species in response to each other's adaptations. Over millions of years, plants and their pollinators have developed specialized traits that benefit both parties. Pollinators, which include a wide range of animals (such as insects, birds, bats, and even wind and water), help plants reproduce by transferring pollen from one flower to another, facilitating fertilization. In turn, plants provide pollinators with food sources, such as nectar and pollen. This mutually beneficial relationship has led to a variety of adaptations in both plants and pollinators.

The Dynamics of Co-evolution

The co-evolution between plants and their pollinators can be seen as a dynamic, ongoing process driven by natural selection. As pollinators evolve to exploit specific plant resources, plants, in turn, evolve to attract and maximize the effectiveness of their pollinators. These interactions result in a variety of morphological, behavioral, and ecological adaptations that enhance pollination efficiency.

Plant Adaptations to Attract Pollinators

Plants have developed numerous adaptations to attract specific pollinators. These adaptations enhance the likelihood that pollinators will visit a flower, increasing the chances of successful pollination. Key plant adaptations include:

- **Flower Color and Shape:** The color and shape of flowers are often tailored to attract particular pollinators. For example:
  - Bees are attracted to flowers that are blue or purple, as they can see these colors well. Many bee-pollinated flowers have a simple, open shape that allows easy access to nectar and pollen.
  - Bird-pollinated flowers, such as those pollinated by hummingbirds, tend to be red or orange, colors that are visible to birds but not to insects. These flowers often have tubular shapes to accommodate the birds' long beaks and tongues.
  - Butterflies are attracted to pink, purple, or orange flowers with flat surfaces to rest on while feeding.
  - Bats prefer flowers that are large, pale, and aromatic, often with a strong smell to attract nocturnal pollinators.
- **Nectar and Pollen:** Flowers often produce nectar as a reward for pollinators. Nectar is a sugar-rich substance that provides energy, while pollen contains proteins and other nutrients. These resources serve as an incentive for pollinators to visit multiple flowers. Plants often fine-tune the amount of nectar and pollen produced based on the type of pollinator they attract. For example:
  - Bee-pollinated flowers often produce large amounts of both nectar and pollen.

- Bird-pollinated flowers tend to have copious amounts of nectar, but less pollen.

- **Odor and Scent:** The scent emitted by flowers plays a crucial role in attracting pollinators. For example:
  - Night-blooming flowers that are pollinated by moths and bats often emit strong, musky odors to attract their nocturnal visitors.
  - Flowers pollinated by bees emit sweet or fruity fragrances.
  - Carrion flowers, like those of the genus *Amorphophallus*, produce foul odors resembling rotting flesh to attract flies and beetles.

- **Flowering Time:** The timing of when a plant blooms can also be adapted to the activity patterns of its primary pollinators. For example:
  - Flowers pollinated by bees tend to open during the day when bees are most active.
  - Flowers pollinated by bats often bloom at night when bats are foraging.

Pollinator Adaptations to Facilitate Pollination

Pollinators have evolved a range of adaptations that enable them to find, access, and feed on plant resources more efficiently, while simultaneously transferring pollen between flowers. Key pollinator adaptations include:

- **Morphological Specializations:** Pollinators have evolved specialized body structures that allow them to exploit certain plant features. Some examples include:
  - **Hummingbird bills:** Hummingbirds have long, slender bills that allow them to feed on the nectar of long-tubed flowers. Their tongues are also specialized for reaching deep into flowers to collect nectar.
  - **Bee body hairs:** Bees have hair-covered bodies that help them collect pollen as they move from flower to flower. The structure of their bodies and legs is adapted for pollen transport, making them highly effective pollinators.
  - **Butterfly proboscis:** Butterflies have long, coiled proboscises that allow them to feed from deep flowers.

- **Behavioral Specializations:** Pollinators also exhibit behavioral adaptations that enhance pollination efficiency, such as:
  - **Flower constancy:** Many pollinators, especially bees, show flower constancy, which means they visit flowers of the same species in succession, increasing the chances of cross-pollination.
  - **Buzz pollination:** Some bees have evolved the ability to perform buzz pollination, where they vibrate their bodies at a specific frequency to release pollen from certain flowers, such as tomatoes or blueberries.

☆ **Co-evolutionary Interactions:** Over time, pollinators and plants may become highly specialized to each other. For example:

❖ **Hummingbirds and long-tubed flowers** have co-evolved such that the shape of the flower matches the bird's bill, and the bird is adapted to hover and feed from the flowers without perching.

❖ **Bees and orchids**: Some orchids have evolved intricate mechanisms to ensure that they are pollinated by specific bee species. For example, certain orchids mimic the appearance and scent of female bees to attract male bees for pollination.

Reciprocal Adaptations and Fitness Benefits

Co-evolution between plants and pollinators leads to reciprocal adaptations that increase the reproductive success (fitness) of both partners. These adaptations enhance both the pollination efficiency of the plant and the foraging efficiency of the pollinator.

☆ **Increased Pollination Efficiency:** As plants evolve to attract specific pollinators, they become more effective at ensuring that their pollen is transferred between flowers, which increases their chances of successful reproduction. In turn, plants may evolve to reward pollinators with more or higher-quality nectar or pollen, incentivizing pollinators to continue visiting.

☆ **Increased Foraging Efficiency:** Pollinators that specialize in certain plant species may benefit from more efficient foraging. By focusing on specific plant species, pollinators can maximize their energy intake, as these plants provide a consistent and reliable food source.

☆ **Specialization vs. Generalization:** While some plants and pollinators evolve to be highly specialized, others remain generalists. Generalists, such as some species of bees or birds, visit a wider variety of plants and can help pollinate many different species. However, specialist interactions often lead to more efficient pollination for both partners.

Examples of Co-evolution Between Plants and Pollinators

☆ **Bees and Flowers:** Many flowers are specifically adapted to attract bees through their color, shape, and scent. For example, flowers like lavender and sunflowers are rich in nectar and attract bees. In return, bees have evolved to efficiently collect pollen from flowers while feeding on nectar. Their hairy bodies are ideal for pollen transfer, which is crucial for the fertilization of flowers.

☆ **Hummingbirds and Red Tubular Flowers:** Hummingbirds are often the primary pollinators for plants with bright red, tubular flowers. The shape of the flower is ideal for the bird's long bill and hovering flight style. Hummingbirds feed on the nectar and, in the process, transfer pollen from one flower to another. In turn, hummingbirds have evolved long bills and tongues that allow them to extract nectar from deep flowers while avoiding excessive water loss.

- ☆ **Bees and Orchids (Sexual Deception):** Some orchids, such as those in the genus Ophrys, use sexual deception to attract their pollinators. These orchids mimic the appearance and scent of female bees, tricking male bees into attempting to mate with the flowers. In the process, the male bee inadvertently transfers pollen to the flower. This strategy ensures pollination without the orchid needing to provide nectar or other rewards.
- ☆ **Bats and Night-Blooming Flowers**: Bats, particularly in tropical and desert ecosystems, are important pollinators for plants with large, fragrant, and pale-colored flowers that bloom at night. Bats are attracted to the scent and size of the flowers and feed on the nectar. While feeding, bats transfer pollen, enabling plants like agave and cactus to reproduce.

The Complexity of Co-evolution

The co-evolution of plants and pollinators is a fascinating and dynamic process that involves mutual adaptations over time. These interactions not only enhance the reproductive success of plants but also provide pollinators with essential food sources. The specialized traits that arise from this co-evolution are crucial for maintaining biodiversity, as the successful reproduction of many plant species depends on their relationship with specific pollinators. In turn, the availability of food resources for pollinators is tightly linked to the reproductive success of plants. Thus, the co-evolution of plants and pollinators is an essential component of ecological processes, contributing to the health and stability of ecosystems.

Chapter 12

Practical Applications of Plant Physiology

Introduction

Plant physiology, the study of the functions and processes occurring within plants, is a cornerstone of understanding how plants interact with their environment and survive in diverse ecosystems. The practical applications of plant physiology extend far beyond the confines of basic science, influencing agriculture, environmental management, biotechnology, and even medicine. By unraveling the complexities of processes such as photosynthesis, transpiration, nutrient uptake, and hormonal regulation, researchers and practitioners can develop innovative strategies to optimize crop production, conserve water, enhance plant resistance to diseases and pests, and improve food security. Furthermore, plant physiological principles underpin the development of sustainable agricultural practices, climate change mitigation efforts, and the creation of bio-based products such as biofuels, pharmaceuticals, and biodegradable materials. In an era of environmental uncertainty and rapid technological advancement, plant physiology has never been more crucial in shaping solutions for global challenges, making it a dynamic field with wide-ranging implications for both the natural world and human society.

Crop Improvement and Yield Enhancement

Crop improvement and yield enhancement are fundamental for ensuring global food security, especially in the face of growing environmental and socio-economic challenges. By combining genetic advances with improved agricultural practices, we can develop crops that not only yield more but are also more resilient, nutritious, and environmentally sustainable. As scientific advancements continue, the potential to revolutionize agriculture and meet the needs of a growing population becomes ever more achievable. Crop improvement and yield enhancement are key goals in agricultural science, particularly in the context of meeting the food and resource demands of a growing global population. These two interrelated areas involve optimizing plant characteristics to improve crop productivity, resilience, and nutritional value, while also addressing environmental challenges. The primary aim is to increase agricultural output, reduce losses, and ensure that crops can thrive under diverse and often changing environmental conditions.

Crop Improvement: Definition and Methods

Crop improvement refers to the genetic modification or selection of plants to enhance desirable traits such as higher yield, disease resistance, improved nutritional content, or adaptability to specific environmental conditions. There are several methods used for crop improvement:

- **Traditional Breeding:** This involves the selection and cross-breeding of plants with desirable traits over multiple generations. Breeders may select plants with traits like larger fruit size, better drought resistance, or disease resistance. The goal is to develop new cultivars with improved characteristics that suit local growing conditions.
- **Hybridization:** Hybrid crops are produced by crossing two genetically different plants to produce offspring with superior traits (*e.g.*, higher yield, disease resistance). Hybrids often exhibit "hybrid vigor" (heterosis), where the offspring show better performance than their parents in certain traits.
- **Genetic Modification (GMOs):** Modern biotechnology techniques allow the direct alteration of a plant's DNA. Genes from other organisms (including bacteria, fungi, or other plants) can be introduced into a crop plant to impart desirable traits, such as resistance to pests, tolerance to herbicides, or improved nutritional content. GMOs have been particularly important in crops like Bt cotton (which resists insect pests) and Roundup Ready soybeans (which tolerate herbicide use).
- **Gene Editing (CRISPR/Cas9):** This advanced technique allows precise alterations to a plant's genome, offering the potential for more specific modifications than traditional genetic modification methods. CRISPR technology is being used to create plants with enhanced stress tolerance, disease resistance, and improved yield.
- **Mutation Breeding:** Using chemicals or radiation to induce mutations in plants, this method produces genetic diversity that may result in the emergence of plants with valuable traits, such as resistance to diseases, pests, or environmental stresses.

Yield Enhancement: Definition and Methods

Yield enhancement focuses specifically on increasing the amount of usable crop harvested per unit area, which is critical for ensuring food security in a world where arable land is limited. Several strategies can enhance crop yields:

- **Nutrient Management:** Proper fertilization is crucial for maximizing crop growth. Nutrient management techniques involve applying the correct amount and type of fertilizers to ensure that plants receive adequate nutrients like nitrogen, phosphorus, and potassium. Advances in plant physiology have also led to the development of more efficient fertilizers that reduce wastage and minimize environmental impact.
- **Water Management and Irrigation:** Efficient water use is critical for maximizing yields, especially in regions facing water scarcity. Techniques such as drip irrigation, soil moisture sensors, and drought-resistant crop varieties help maintain optimal growing conditions while conserving water. Additionally, research in plant physiology has led to the development of drought-tolerant crops that are better able to survive with minimal water.
- **Pest and Disease Management:** Pests and diseases can significantly reduce crop yields. Crop improvement through breeding for resistance to insects, fungi, and viruses is a key method of reducing yield loss. Integrated pest management (IPM) combines biological control (using natural predators of pests), chemical control (targeted pesticide application), and cultural practices (like crop rotation) to minimize pest damage and maintain yields.
- **Improved Photosynthesis and Growth Efficiency:** One of the primary drivers of yield enhancement is improving photosynthetic efficiency. Plants with higher photosynthetic rates can convert sunlight into energy more efficiently, which translates into higher biomass and ultimately higher yields. Advances in plant physiology have led to the identification of genetic pathways that regulate photosynthesis, and researchers are working to enhance this process in crops.
- **Stress Tolerance:** Environmental stress, such as drought, high temperatures, soil salinity, and extreme weather events, can severely reduce crop yields. To enhance crop yield under these conditions, scientists are developing varieties that are more resilient to stress through genetic modification, breeding, and physiological studies. Stress-tolerant crops are able to maintain normal growth and reproduction even in suboptimal conditions, leading to more stable and reliable yields.
- **Plant Growth Regulators (PGRs):** PGRs are chemicals that influence plant growth and development. The application of growth regulators can help improve seed germination, promote root development, increase fruit set, or delay senescence, all of which can enhance yield. For example,

gibberellins are used to promote seedling growth, while auxins may be used to increase fruit size.

Role of Molecular Biology and Biotechnology in Yield Enhancement

Molecular biology tools play an integral role in enhancing crop yield by enabling the precise identification of genes responsible for key traits, such as drought tolerance, disease resistance, and nutrient use efficiency. Some of the techniques used include:

- **Marker-Assisted Selection (MAS):** This technique uses molecular markers to identify desirable genes or traits in plants, accelerating the breeding process. MAS has been used to develop varieties of crops that are resistant to diseases like rust or blight, or that have improved nutrient profiles.
- **Genetic Transformation:** Transgenic crops, or genetically modified organisms (GMOs), are developed by inserting specific genes that confer traits like resistance to pests (*e.g.*, Bt cotton) or tolerance to abiotic stresses. This direct manipulation of the genetic makeup of plants is a powerful tool for enhancing yield and improving crop sustainability.
- **Metabolic Engineering:** This involves modifying the metabolic pathways of plants to improve nutrient utilization or enhance the synthesis of compounds that improve yield or quality. For example, genetically engineered crops may produce higher levels of essential amino acids or vitamins, which can lead to more nutritious food crops.

Climate Change and Yield Enhancement

As climate change continues to impact global agriculture, crop improvement and yield enhancement strategies are essential for ensuring food security. With rising temperatures, altered precipitation patterns, and more frequent extreme weather events, crops must be able to adapt to new challenges.

- **Climate-Resilient Crops:** Developing crops that can withstand the changing climate is a critical area of research. This includes breeding or engineering plants with enhanced resistance to heat, drought, flooding, and salinity.
- **Carbon Dioxide Enrichment:** Elevated levels of CO_2 in the atmosphere can affect photosynthesis and growth. Some crops may respond positively to higher CO_2 concentrations by increasing photosynthetic efficiency, potentially enhancing yields. Research into how different crops respond to elevated CO_2 will be important for adapting to a future with higher atmospheric carbon levels.

Integrated Approaches to Crop Improvement and Yield Enhancement

The most effective way to increase crop yield and improve plant characteristics is through an integrated approach that combines traditional breeding with modern biotechnological tools, along with sustainable agricultural practices. This holistic approach can address the many challenges facing modern agriculture, such as limited land, climate change, and the need for food security.

For example, an integrated approach might involve:

- Developing high-yielding, stress-tolerant crops through genetic engineering.
- Optimizing agronomic practices (fertilization, irrigation, pest control).
- Utilizing smart technologies (*e.g.*, drones, sensors) to monitor plant health and optimize input use.
- Promoting sustainable practices like reduced tillage, crop rotation, and agroecological approaches to enhance soil health.

Plant Physiology in Horticulture and Physiology

Plant physiology plays an indispensable role in horticulture by providing a deep understanding of plant processes that can be harnessed to improve crop productivity, quality, and sustainability. Whether through optimizing photosynthesis, managing water and nutrients, applying growth regulators, or controlling post-harvest quality, knowledge of plant physiology empowers horticulturists to make informed decisions that enhance the growth, yield, and resilience of horticultural crops. With advancements in molecular biology, biotechnology, and environmental management, the integration of plant physiology into horticultural practices continues to drive innovations that meet the challenges of food security and sustainable agriculture. Plant physiology is a central science in horticulture, the branch of agriculture focused on the cultivation of fruits, vegetables, nuts, seeds, herbs, sprouts, mushrooms, algae, and non-food crops like grass and flowers. Understanding the physiological processes that govern plant growth, development, and response to environmental conditions is essential for improving horticultural practices, optimizing crop yields, and ensuring the sustainability and quality of horticultural products. In horticulture, plant physiology directly informs decisions related to plant care, cultivation, breeding, pest management, and post-harvest handling.

Photosynthesis and Carbon Assimilation in Horticultural Crops

Photosynthesis is the primary process through which plants convert sunlight into chemical energy, a vital aspect of all plant growth and productivity. In horticulture, maximizing photosynthetic efficiency is a key strategy for improving crop yields, especially in fruit and vegetable production.

- ☆ **Optimizing Light Use Efficiency:** Understanding how light is captured and converted into plant biomass allows horticulturists to optimize growing conditions. This includes choosing the right lighting for greenhouse-grown crops or managing canopy structure for open-field crops to maximize light interception.
- ☆ **Photosynthetic Pathways in C3, C4, and CAM Plants:** Different plants use different photosynthetic pathways. For instance, C4 plants (like maize and sugarcane) have higher efficiency in hot, sunny environments, while CAM (Crassulacean Acid Metabolism) plants (such as succulents and cacti) are adapted to arid conditions. Horticulturists select crop varieties based on these adaptations to optimize yield in specific climates.

Water Relations and Irrigation Management

Water management is critical in horticulture, where water scarcity or excess can have a direct impact on crop health and yield. Understanding plant water relations – the processes by which plants absorb, transport, and lose water – is essential for managing irrigation systems and ensuring crops receive optimal moisture.

- ☆ **Transpiration and Water Loss:** Transpiration, the process by which water is lost through the stomata of leaves, is linked to plant cooling and nutrient transport. However, excessive transpiration can lead to water stress, especially in fruit-bearing plants. Efficient irrigation methods, such as drip or micro-sprinkler systems, are employed to minimize water loss and ensure consistent moisture availability.
- ☆ **Drought Resistance Mechanisms:** In water-scarce regions, developing drought-resistant cultivars or utilizing irrigation strategies that optimize water use efficiency is crucial. Plants can exhibit various physiological adaptations, such as stomatal closure, increased root depth, or altered metabolic pathways to reduce water loss and maintain productivity under stress.

Nutrient Uptake and Fertilization

Plants require a range of nutrients for growth, with essential macro-nutrients (nitrogen, phosphorus, potassium, calcium, magnesium, and sulfur) and micro-nutrients (iron, manganese, zinc, copper, boron, molybdenum, and chlorine) playing key roles in physiological processes like photosynthesis, energy production, and cell division.

- ☆ **Nutrient Absorption and Transport:** Plants take up nutrients from the soil via their roots and transport them through vascular tissues (xylem and phloem). A balanced supply of nutrients is crucial for optimizing plant growth, fruit development, and resistance to diseases. Nutrient deficiencies or imbalances can lead to poor growth, reduced yields, and susceptibility to environmental stresses.

- **Fertilizer Management:** Understanding plant nutrient requirements is essential for effective fertilizer application. Over-fertilization can lead to nutrient runoff, water pollution, and soil degradation, while under-fertilization can limit crop productivity. Fertilizer management strategies, including organic amendments and controlled-release fertilizers, can optimize nutrient uptake and enhance plant health.

Growth Regulators and Developmental Physiology

Plant growth regulators (PGRs) are naturally occurring or synthetic compounds that influence a wide range of plant processes, from seed germination to flowering, fruit set, and senescence. In horticulture, manipulating these growth regulators can significantly enhance crop quality, uniformity, and yield.

- **Auxins and Root Development:** Auxins, particularly indole-3-acetic acid (IAA), play a crucial role in cell elongation, root formation, and branching. In horticulture, auxin applications are used to promote rooting in cuttings, enhance root systems in transplants, and regulate growth in various crops.
- **Cytokinins and Shoot Development:** Cytokinins promote cell division and are involved in shoot initiation and development. In horticultural practices, cytokinins are used to encourage branching in ornamental plants or stimulate flower bud development in fruit trees.
- **Ethylene and Fruit Ripening:** Ethylene is a gas that plays a central role in the ripening of fruits. In commercial horticulture, understanding ethylene production allows growers to control fruit ripening, extending shelf life and improving quality. For example, controlled atmospheric storage is used to delay ripening in fruits like apples, bananas, and tomatoes.

Reproductive Physiology in Horticultural Crops

Reproductive processes are central to horticulture because they determine yield and fruit quality. Understanding the physiological aspects of pollination, fertilization, and seed development helps growers improve the reproductive success of their crops.

- **Pollination and Fertility:** In many crops, cross-pollination by insects, birds, or wind is necessary for fertilization and fruit set. Some crops, like tomatoes and peppers, are self-pollinating, while others require controlled pollination. Physiological factors like flower morphology, pollen viability, and environmental conditions all influence pollination success.
- **Flowering Regulation:** Flowering is a critical stage in crop production, especially for fruit-bearing plants. Horticulturists use plant physiology knowledge to manipulate flowering timing for optimal harvests, often using environmental cues like temperature (vernalization), light

(photoperiod), and growth regulators (*e.g.*, gibberellins) to trigger or delay flowering.

Post-Harvest Physiology and Quality Maintenance

Post-harvest handling and storage are critical aspects of horticultural practices, as they directly affect the shelf life, quality, and marketability of crops. Understanding the physiological processes that occur after harvesting can help preserve the quality of fruits, vegetables, and flowers.

- ☆ **Respiration and Ripening:** After harvest, fruits and vegetables continue to respire, consuming oxygen and producing carbon dioxide. This metabolic activity is linked to ripening, senescence, and spoilage. By controlling temperature, humidity, and gas composition (*e.g.*, using controlled or modified atmosphere storage), horticulturists can slow respiration, delay ripening, and extend the shelf life of perishable products.
- ☆ **Ethylene Control:** As mentioned earlier, ethylene is a key regulator of fruit ripening. By controlling ethylene exposure, growers can extend storage times and control the timing of fruit ripening, ensuring that fruits reach optimal quality at the right time for market consumption.

Stress Physiology in Horticultural Crops

Horticultural crops are often subject to abiotic stresses such as drought, heat, salinity, and frost, as well as biotic stresses like diseases and pests. Stress physiology examines how plants respond to these challenges and how horticulturists can mitigate stress impacts.

- ☆ **Heat and Cold Stress:** Temperature extremes can damage cellular structures, enzymes, and metabolic processes. Research into heat and cold tolerance mechanisms allows horticulturists to breed or select cultivars that can thrive in specific climates or adapt to changing weather conditions.
- ☆ **Salt Tolerance:** In regions where soil salinity is high, developing salt-tolerant crops is a priority. Plants exhibit various mechanisms, such as the exclusion of salt from roots or the compartmentalization of salt in vacuoles, to survive saline conditions.
- ☆ **Disease Resistance:** Understanding the physiological interactions between plants and pathogens is key to developing disease-resistant varieties. Plant resistance can be due to physical barriers (*e.g.*, thickened cell walls) or biochemical responses (*e.g.*, the production of defensive proteins or secondary metabolites).

Managing Water Use Efficiency in Crops

Water is one of the most critical resources for agricultural production, and its efficient use has become increasingly vital in the face of climate change, population growth, and the rising demand for food. Managing water use efficiency (WUE) in crops involves maximizing crop productivity per unit of water used, ensuring that plants receive the optimal amount of water to achieve high yields without wasting or depleting valuable water resources. With agriculture being the largest consumer of freshwater worldwide, improving water use efficiency is not only essential for increasing crop productivity but also for ensuring the sustainability of agricultural systems and mitigating the environmental impacts associated with excessive water usage.

Understanding Water Use Efficiency (WUE)

Water use efficiency in agriculture refers to the ratio of crop yield (usually in terms of biomass or economic yield) to the amount of water used during the growing season. The concept can be broken down into two main components: water productivity (the amount of crop produced per unit of water used) and water use efficiency (how effectively water is used by the plant in relation to the environmental conditions). WUE can be influenced by a variety of factors, including crop type, soil properties, climate conditions, irrigation practices, and crop management techniques.

There are two key aspects of WUE:

- ✰ **Biological WUE:** This focuses on the efficiency of photosynthesis, transpiration, and other physiological processes that drive plant growth. It reflects the plant's ability to convert water into biomass and yield.
- ✰ **Agronomic WUE:** This relates to the practices used by farmers to optimize water use, such as irrigation management, soil fertility, and crop management practices. It also includes the water used for soil preparation and other farm operations.

Improving WUE in crops not only leads to higher yields but also reduces water consumption, making agricultural systems more sustainable, especially in regions experiencing water scarcity or erratic rainfall.

Strategies for Enhancing Water Use Efficiency in Crops

Several strategies can be implemented to manage and improve water use efficiency in crop production. These strategies range from genetic improvements and crop management practices to technological interventions in irrigation systems.

Genetic Improvements and Breeding for Water Efficiency

One of the most promising approaches for improving water use efficiency is through the development of crop varieties that are genetically adapted to use water more efficiently. Advances in plant breeding and genetic engineering have led to the development of drought-resistant crops, which exhibit enhanced water use efficiency by making better use of available moisture.

- **Drought Tolerance Mechanisms:** Breeding crops that can maintain high yields under water-limited conditions is a key area of focus. Drought-tolerant crops have mechanisms that allow them to conserve water, such as deeper root systems that access moisture from lower soil layers or the ability to close stomata to reduce water loss through transpiration. Additionally, some crops can alter their metabolic processes to better tolerate water stress, such as by maintaining higher levels of osmotic regulators like proline.
- **Root System Development:** Crops with deep or more extensive root systems are better equipped to extract water from the soil, particularly in dry periods. Advances in root architecture research aim to develop varieties with enhanced root systems that can explore larger soil volumes, increasing their access to water during drought conditions.

Efficient Irrigation Practices

Irrigation is a major component of water use in agriculture, particularly in regions with insufficient rainfall. Optimizing irrigation practices is critical to improving water use efficiency and ensuring that water is applied where and when it is needed most.

- **Drip Irrigation:** Drip irrigation delivers water directly to the plant roots, minimizing evaporation losses and runoff. This system allows water to be applied in precise amounts, improving water use efficiency by ensuring that crops receive water in a timely and controlled manner. Drip irrigation has proven particularly effective for high-value crops like vegetables and fruits, where optimal water management is crucial to maximize yield and quality.
- **Micro-Sprinklers and Subsurface Irrigation:** These methods provide uniform water distribution while reducing water wastage. Micro-sprinklers use small nozzles to disperse water in a fine mist, while subsurface irrigation applies water directly to the root zone, minimizing evaporation and surface runoff. Both methods have been shown to reduce water usage compared to traditional flood irrigation.
- **Irrigation Scheduling and Monitoring:** Advanced irrigation management techniques, such as soil moisture sensors and weather-based irrigation controllers, help determine the exact water requirements of crops. These tools ensure that water is applied only when needed, preventing over-irrigation and under-irrigation. Automated irrigation systems that adjust watering schedules based on real-time weather data and soil moisture content can significantly enhance water use efficiency.

Water-Conserving Agricultural Practices

In addition to irrigation technologies, there are several agronomic practices that can improve water use efficiency in crops. These include soil management, mulching, and crop selection.

- ☆ **Soil Health and Conservation Tillage:** The soil plays a crucial role in water retention. Healthy soils with good organic matter content retain moisture better and support deeper root growth. Practices like conservation tillage, which minimizes soil disturbance, help maintain soil structure and improve its ability to retain water. Reduced tillage also reduces soil evaporation and runoff, further enhancing water use efficiency.
- ☆ **Cover Cropping and Mulching:** Mulching with organic materials, such as straw or compost, helps retain soil moisture by reducing evaporation. Additionally, cover crops can reduce soil erosion, improve water infiltration, and enhance soil organic matter, all of which contribute to better water retention and more efficient water use.
- ☆ **Crop Rotation and Intercropping:** These practices help improve soil health and optimize water use by growing crops with different water requirements or planting drought-tolerant species alongside more water-demanding crops. These systems can reduce competition for water and improve overall water efficiency across the farm.

Climate-Smart Agricultural Practices

With the increasing frequency and intensity of extreme weather events due to climate change, climate-smart agricultural practices are essential for managing water use efficiency. These practices are designed to help farmers adapt to changing weather patterns, conserve water, and reduce the impact of water stress on crops.

- ☆ **Climate-Resilient Crop Varieties:** In areas prone to drought or unpredictable rainfall, it is crucial to select crop varieties that are better suited to the local climate. Breeding or selecting drought-tolerant crops that require less water or can withstand periods of water scarcity can help ensure reliable yields even in challenging conditions.
- ☆ **Rainwater Harvesting:** In regions where rainfall is seasonal or erratic, collecting and storing rainwater for irrigation can be an effective strategy. Rainwater harvesting systems, which capture runoff from rooftops or other surfaces, allow farmers to store water for use during dry periods, reducing their reliance on groundwater or expensive irrigation systems.

Technological Innovations for Water Management

Technological advancements play a critical role in improving water use efficiency in crop production. Precision agriculture, which uses data-driven technologies to monitor and manage agricultural practices, has the potential to revolutionize water management on farms.

- **Remote Sensing and Satellite Technology:** Remote sensing technologies, such as satellite imagery and drones, allow farmers to monitor crop health, water stress, and soil moisture content across large areas. This data can be used to make informed decisions about irrigation schedules and crop management practices, improving water efficiency and reducing waste.
- **Data-Driven Irrigation Systems:** The integration of Internet of Things (IoT) sensors and real-time data collection enables farmers to monitor soil moisture levels, weather conditions, and crop water needs. This technology allows for precise irrigation control, which optimizes water use and minimizes waste.

Policy and Economic Considerations

To achieve long-term improvements in water use efficiency, governments and policy-makers must create a supportive framework that encourages the adoption of water-efficient technologies and practices. Policies that promote the efficient use of water in agriculture include:

- **Incentives for Water-Efficient Technologies:** Governments can provide subsidies or tax incentives to encourage farmers to adopt water-efficient irrigation systems, such as drip irrigation or precision irrigation technologies. These incentives help offset the initial investment costs, making it easier for farmers to transition to more sustainable practices.
- **Water Pricing and Allocation:** Fair and transparent water pricing policies can incentivize efficient water use while ensuring that water resources are allocated equitably. In many regions, water scarcity is exacerbated by inefficient water pricing, which can lead to over-extraction and waste. Proper water pricing mechanisms can encourage more responsible water use by farmers and other stakeholders.
- **Research and Extension Services:** Governments and agricultural institutions should invest in research and extension services that promote water-efficient farming practices. Providing farmers with the knowledge and tools to manage water efficiently is crucial for enhancing agricultural sustainability.

Efficient water use is critical for ensuring the sustainability of global food production in the face of increasing water scarcity and the impacts of climate change. By integrating advanced irrigation technologies, selecting drought-tolerant crop varieties, employing soil and crop management practices, and adopting climate-smart agricultural techniques, farmers can significantly enhance water use efficiency. With continued research, innovation, and supportive policy frameworks, water use efficiency can be improved across the agricultural sector, helping to secure food production and protect valuable water resources for future generations.

Plants in Environmental Remediation and Conservation

Plants are indispensable agents in environmental remediation and conservation, offering sustainable, cost-effective solutions to many of the environmental challenges humanity faces today. Whether through cleaning polluted soils and waters, restoring degraded habitats, or sequestering carbon to mitigate climate change, plants provide a wealth of ecosystem services that are essential to the health of the planet. As research and technological advancements continue to unlock the full potential of plant-based solutions, we can expect an increasingly vital role for plants in achieving a more sustainable and resilient world.Plants play a pivotal role in addressing some of the most pressing environmental challenges of our time, such as pollution, habitat loss, soil degradation, and climate change. As primary producers in ecosystems, plants not only contribute to the stabilization of natural systems, but they can also be harnessed in environmental remediation and conservation efforts. The ability of plants to improve environmental quality–whether by absorbing toxic substances, stabilizing soils, or restoring degraded habitats–has opened up a range of innovative solutions for sustainable environmental management. From phytoremediation to carbon sequestration, plants offer a natural, low-cost, and often highly effective means of mitigating environmental damage and fostering ecological balance.

Phytoremediation: Harnessing Plants for Pollution Cleanup

Phytoremediation is the process of using plants to remove, degrade, or immobilize pollutants from soil, water, and air. This approach capitalizes on the natural abilities of certain plants to accumulate, detoxify, or transform harmful substances. Compared to traditional remediation methods, phytoremediation is often more cost-effective, eco-friendly, and sustainable. Some common forms of phytoremediation include:

Phytoextraction

In this process, plants absorb heavy metals or other pollutants from the soil into their tissues. Plants like Indian mustard (*Brassica juncea*) and sunflower (*Helianthus annuus*) have been shown to accumulate metals such as lead (Pb), cadmium (Cd), and arsenic (As) from contaminated soils. These plants are typically harvested once they have taken up high levels of the contaminants, thus removing the toxic substances from the environment.

Applications: Phytoextraction has been applied to clean up contaminated industrial sites, mining areas, and urban soils with high levels of toxic metals. It can also be used in agricultural fields that have been exposed to excessive use of pesticides or other chemicals.

Phytodegradation

Phytodegradation involves the breakdown of organic pollutants through enzymatic processes in plants. Certain plants can metabolize pollutants such as pesticides, solvents, and petroleum products in their tissues, converting them into less harmful substances.

Applications: This method is particularly useful for dealing with organic contaminants such as hydrocarbons (oil spills) and herbicides. For example, poplar trees (*Populus* spp.) and willows (*Salix* spp.) are known to degrade certain organic contaminants like solvents and petroleum products.

Rhizofiltration

Rhizofiltration refers to the process by which plant roots absorb or filter out contaminants from water. This method is especially useful for removing heavy metals, radionuclides, and organic pollutants from contaminated water bodies, including groundwater, rivers, and lakes. The roots of certain plants such as water hyacinth (*Eichhornia crassipes*) and duckweed (*Lemna* spp.) can filter and absorb toxins from polluted water, offering an eco-friendly alternative to conventional water treatment methods.

Applications: Rhizofiltration has been applied in cleaning up industrial wastewater, agricultural runoff, and urban stormwater. It has also been used in aquaculture systems to help remove excess nutrients and pollutants.

Phytostabilization

Phytostabilization involves the use of plants to immobilize contaminants in the soil, preventing their spread and reducing the risk of contamination in the surrounding environment. This is typically achieved by root systems that bind pollutants in the soil, reducing their mobility and leaching into groundwater. Plants like grass species, clover (*Trifolium* spp.), and willows are effective at stabilizing contaminated soils, especially in mining and industrial sites.

Applications: Phytostabilization is often used in the remediation of areas where heavy metal contamination is present, such as old mining sites or industrial zones, to prevent further contamination of groundwater and reduce risks to human health and local wildlife.

Restoring Degraded Ecosystems and Habitat Conservation

Plants also play a critical role in habitat conservation and the restoration of ecosystems that have been damaged by human activity or natural disasters. Their ability to stabilize soils, rebuild biodiversity, and restore ecological balance is central to ecosystem restoration efforts.

Erosion Control and Soil Stabilization

Plants are key players in controlling soil erosion, which can be caused by deforestation, overgrazing, agricultural practices, or natural events like floods and hurricanes. Deep-rooted plants and grasses, such as vetiver grass (*Chrysopogon zizanioides*), buffalo grass (*Buchloe dactyloides*), and clover, are often used to stabilize soil and prevent erosion on slopes, riverbanks, and coastal areas.

Applications: Erosion control and soil stabilization using plants have been widely employed in reforesting degraded landscapes, reclaiming mining sites, restoring riverbanks, and protecting coastal zones from storm surges.

Habitat Restoration and Biodiversity Conservation

Restoring native plant species in areas that have been altered by agriculture, urbanization, or industrial activities can help revive lost biodiversity and improve ecosystem services. By reintroducing native plants, ecosystems can recover their natural functions, such as pollination, carbon sequestration, and water regulation.

Applications: In areas that have suffered from deforestation or agricultural expansion, ecological restoration efforts often involve planting native tree species, shrubs, and grasses. For example, the restoration of tropical rainforests and wetlands often begins with planting species that are native to the region, helping to restore soil fertility and create suitable habitats for wildlife.

Carbon Sequestration

Plants play a vital role in mitigating climate change by capturing and storing carbon dioxide (CO_2) from the atmosphere. This process, known as carbon sequestration, is carried out through photosynthesis, where plants absorb CO_2 and convert it into biomass (*e.g.*, roots, stems, and leaves). Forests, grasslands, and wetlands are particularly important in carbon sequestration, as they act as long-term carbon sinks, holding large amounts of carbon in their soil and vegetation.

Applications: Large-scale reforestation projects, afforestation (planting forests in areas that were not previously forested), and the restoration of degraded grasslands and wetlands are all strategies used to increase carbon sequestration and mitigate the impacts of climate change.

Wetland Restoration and Water Purification

Wetlands are among the most valuable ecosystems on the planet, offering numerous ecological services such as water purification, habitat for wildlife, flood regulation, and carbon storage. However, wetland ecosystems have been severely impacted by urbanization, agriculture, and pollution. Restoration of these ecosystems involves the planting of native wetland plants, which help filter pollutants, stabilize sediments, and promote biodiversity.

Applications: Reintroducing wetland plants such as bulrush (Schoenoplectus spp.), cattail (*Typha* spp.), and reeds (*Phragmites australis*) is a common approach in wetland restoration efforts. These plants help purify water by absorbing excess nutrients, heavy metals, and organic pollutants, and they also provide crucial habitat for aquatic species.

The Role of Plants in Climate Change Mitigation

In addition to carbon sequestration, plants play a broader role in climate change mitigation and adaptation. They are fundamental to maintaining ecological balance and regulating atmospheric gases, particularly in managing the effects of greenhouse gas emissions.

- ☆ **Green Urban Infrastructure:** Cities are increasingly turning to green infrastructure, such as green roofs, urban forests, and plant-covered walls, to mitigate the urban heat island effect, reduce carbon footprints, and improve air quality. These urban plantings help absorb CO_2, provide shade, and enhance biodiversity in urban areas.
- ☆ **Agroforestry:** The integration of trees and other vegetation into agricultural systems (agroforestry) provides multiple benefits, including carbon sequestration, improved soil health, and enhanced biodiversity. Agroforestry can also help mitigate the impact of climate change on crop yields by improving water retention and reducing the need for synthetic inputs.

Challenges and Future Directions

While the potential of plants in environmental remediation and conservation is vast, there are challenges to their widespread application. These challenges include the long time frames required for some remediation processes, the need for suitable plant species for specific environments, and the risks associated with the accumulation of pollutants in plant tissues. Moreover, the success of plant-based solutions is often contingent on the cooperation of local communities, government policies, and adequate funding.Research into plant physiology, genetic engineering, and ecosystem dynamics continues to enhance the effectiveness of plants in environmental management. The use of genetically modified plants designed for enhanced pollutant uptake or drought tolerance could expand the range of environments in which plant-based remediation is effective. Additionally, increasing our understanding of plant-microbe interactions, particularly in phytoremediation, may further improve the ability of plants to clean up contaminated environments.

Glossary

Abscisic Acid (ABA): A plant hormone involved in stress responses, such as drought tolerance, and in regulating seed dormancy and stomatal closure.

Absorption of water: The way in which water from soil enters roots, particularly to the root xylem is called mechanism of water absorption.

Acclimation: The process by which a plant gradually adjusts to changes in its environment, such as temperature or light intensity.

Acetyl Coenzyme A - A small molecule that carries acetyl functional groups in cells. Composed of an acetyl group attached to a coenzyme A molecule. The starting product of the citric acid cycle.

Adaxial: Referring to the upper or inner side of an organ, such as a leaf (the side facing the stem).

Adenosine diphosphate (ADP): Adenine nucleoside with two phosphate molecules of which terminal one is energy rich.

Adenosine monophosphate (AMP): Adenine nucleoside with two phosphate molecules of which terminal one is energy rich. Adenosine diphosphate (ADP) Adenine nucleoside with two phosphate molecules of which terminal one is energy rich.

Adenosine Triphosphate (ATP): The molecule from which cells derive energy. Comprised of an adenosine molecule bonded to three phosphates, each phosphate bond contains energy, especially the third bond. By breaking that one bond and reducing ATP to adenosine diphosphate (ADP), the cell can get the energy to carry out its various processes.

Adenosine: A nucleoside containing a heterocyclic nitrogen base (adenine) and a pentose sugar.

Aerobic respiration: Aerobic respiration is an enzymatically controlled release of energy in a stepwise catabolic process of complete oxidation of organic food into carbon dioxide and water with oxygen acting as terminal oxidant. The common mechanism of aerobic respiration is also called common pathway because its first step, called glycolysis, is common to both aerobic and anaerobic modes of respiration. The common aerobic respiration consists of three steps– glycolysis, Krebs cycle and terminal oxidation.

Ageing: The process of becoming older.

Alkaloids: The group of naturally occurring chemical compounds that mostly contain basic nitrogen atoms. An amide: A functional group contains a carbonyl group linked to a nitrogen atom and It is the conjugate base of ammonia ($NH3$).

Amino acid: An organic compound serves as a building block for proteins and contains an amino group and a carboxylic acid group.

Ammonification: The formation of ammonia or its compounds by decomposition of organic matter.

Annual: Refers to a plant that completes its entire life cycle in growing season. Marigolds are examples of an annual plant.

Anthocyanins: Blue and red glycoside pigments (flavonoid) found dissolved in cell sap.

Apoplast: Inside a plant, the apoplast is the space outside the plasma membrane within which material can diffuse freely.

Ascent of sap: The upward conduction of water from the roots to different parts of the plant body is called Ascent of sap.

Assimilation: Uptake and formation of simple foodstuff. Auxin-A group of hormones of plants which is synthetized by growing tips of stems and roots and regulate may aspects of plant growth, *e.g.* IAA.

ATPase: Also called myosine, an enzyme catalyzing TTP, CTP, GTP, UTP and ATP to yield orthophosphate and the corresponding diphosphate.

Autotroph: An organism that can produce its own food from inorganic substances, usually through photosynthesis (*e.g.*, most plants).

Auxin: A group of hormones of plants which is synthetized by growing tips of stems and roots and regulate may aspects of plant growth, *e.g.* IAA.

Biennial: Those plants that sprout from seed and grow through one season, overwinter then flower, go to seed and die in the second growing season.

Bio-fertilizers: Any fertilizer of biological origin, which is a substance contains living microorganisms. Pleomorphic bacteria: Some bacteria having the ability of to alter their shape or size in response to environmental conditions.

Biotin: Vitamin H.

Cambium: A layer of actively dividing cells in plants that is responsible for secondary growth, producing new xylem and phloem cells.

Capillary Water: The water which remains present in the intercellular spaces of soil particles. This water is available to plants, which fills the spaces between thenon- colloidal particles or it forms a thin film around the non- colloidal small particles of soil. This water is present in the capillaries of soil particles thus known as capillary water and one of the most important and significant form of water which is held by capillary forces and absorbed by the plants, or the water that is left in soil along with hygroscopic moisture and water vapours after the gravitational water has drained off is known as capillary water.

Carbohydrate: Compound containing carbon, hydrogen and oxygen with general formula Cn (H_2O)n *e.g.* sugars, starch cellulose.

Carotenoids: Yellow, orange or red pigments, soluble in oil solvents. These pigments are is absent such as carrot roots, floral petal *etc.*

Cell wall: In plant cells the outer layer surrounding the plasma membrane. It consists of cellulose, hemicelluloses and pectic substances.

Chlorophyll: A green pigment in chloroplasts that absorbs light energy during photosynthesis, essential for converting light into chemical energy.

Chlorophyll: Primary pigment used in photosynthesis. Plants contain two main forms of chlorophyll: *a* and *b*. Chlorophyll has a hydrocarbon tail that anchors it to an integral protein in the thylakoid membrane of the chloroplast. Chlorophyll is the source of the green color of plants and certain other autotrophs.

Chloroplast: Organelles in plant cells containing chlorophyll, where photosynthesis occurs. They convert light energy into chemical energy.

Chlorosis-Disease of green plants in which green colour is lost and characteristic yellow colour appears.

Citric acid cycle - Also known as the Kreb's Cycle; a metabolic pathway found in aerobic organisms that oxidizes acetyl coA groups to carbon dioxide and water.

Coenzyme: A molecule that participates in an enzyme-catalyzed reaction and functions to transfer atoms or electrons between itself and various molecules.

Companion cell: Type of cell presents in phloem tissue, associated with sieve cells and provide energy for the process of translocation.

Cotyledon: Energy storage components of a seed that feed the plant before the emergence of its first true leaves.

Cytokinin: A class of plant hormones that promote cell division, delay senescence, and influence shoot and root growth.

Cytoplasm-All the protoplasm of a cell excluding nucleus.

Deciduous: A plant whose parts, particularly leaves, are shed at regular intervals or a given stage of development.

Defoliation: The shedding of leaves, which can be a natural process or a response to environmental stress or herbivory.

Denitrification: Reduction of nitrates or nitrites commonly by bacteria (as in soil) that usually results in the escape of nitrogen into the air.

Dicotyledon (Dicot): A group of angiosperms (flowering plants) that typically have two seed leaves (cotyledons), such as beans and sunflowers.

Diffusion pressure deficit: Net capacity of plant cell for absorbing water. FAD-Flavin adenine dinucleotide.

Diffusion- The process whereby particles of liquid, gases or solids are intermingled as the result of their spontaneous movement caused by them.

Dormancy: This term is most often used when referring to a seed's inability to sprout for a given time after it is produced.

Dormant: To remain inactive when conditions are not suitable. Seeds are often dormant until conditions are right for germination.

Dry Weight: Weight of any plant (or other organism) part after all its water content has been removed by drying.

Eectomycorrhizal fungi: A symbiotic relationship that occurs between a fungal symbionts and the plant roots.

Elimination reaction - A reaction that involves the ejection of a specific group from a molecule, often resulting in the formation of a carbon-carbon double bond.

Endosmosis- In plants diffusion of solvent inside the cell is called endosmosis. Exosmosis- In plant diffusion of solvent outside the cell is called endosmosis.

Endosperm: A tissue in seeds that provides nutrition to the developing embryo, formed during fertilization in angiosperms.

Epicotyl: Portion of the axis of a plant embryo or seedling stern between the cotyledons and the primary leaves.

Ethylene: A gaseous plant hormone that regulates fruit ripening, flower wilting, and leaf abscission.

FAD-Flavin adenine dinucleotide. FMN-Flavin mononuicleotide.

Fertilization: The action or process of fertilizing an egg or a female animal or plant, involving the fusion of male and female gametes to form a zygote.

Field capacity- Field capacity is the amount of soil moisture or water content held in the soil after excess water has drained away and the rate of downward movement has decreased.

Flavanoids: Large group of water-soluble plant pigments constitute antioxidant, antiinflammatory, and antiviral properties that are beneficial to health

Floral Buds: A plant bud that produces only a flower. Morphogenesis: Formation of the structure of an organism or part

Germination: Resumption of active growth in an embryo which results in its emergence from the seed and development of those structures essential to plant development.

Gibberellin: A plant hormone that stimulates growth, seed germination, and flowering, and is involved in elongating stems and leaves.

Glutamine synthetase (GS): An enzyme that plays an essential role in the metabolism of nitrogen by catalyzing the condensation of glutamate and ammonia to form glutamate.

Glycolysis - A metabolic pathway occurring in the cell cytosol that during a series of reactions converts glucose to pyruvate and synthesizes ATP.

GOGAT (Glutamine-2-oxoglutarate amino-transferase): An enzyme manufactures glutamate from glutamine and α-ketoglutarate and also plays a central role in the regulation of nitrogen assimilation in photosynthetic eukaryotes and prokaryotes.

Gravitational Water: The water which reaches deeply into the soil after rains due to gravitational force is called gravitational water. This water is not available to plants and percolates downwards through large pores between the soil particles under the gravity and hence is called gravitational water. It reaches the low water table in the earth.

Gravitropism (Geotropism): A plant's growth response to gravity, with roots typically growing downward (positive gravitropism) and stems growing upward (negative gravitropism).

GTP (Guanosine Tri Phosphate): A nucleotide composed of guanine, ribose, and three linked phosphate groups ADP: An organic compound $C_{10}H_{15}N_5O_{10}P_2$ is converted to ATP for the storage of energy during cell metabolism.

Guard Cells: Specialized cells in the epidermis of leaves that control the opening and closing of stomata, regulating gas exchange and water loss.

Guttation- Exudation of water usually through structures called hydathodes present at the tip of veins of leaves.

Haploid: A cell or organism that has one set of chromosomes, as opposed to diploid, which has two sets. Gametes (sperm and egg) are haploid.

Hormone: Chemical substance that modifies the growth and development of a plant. Inflorescence: The name for the structure on the plant that carries the flower.

Humic Substance: An organic residue of decaying organic matter Nitrite reductase: An enzyme that catalyze the reduction of nitrite.

Hydathodes- Structures present at the ends of vein lets of leaves, specially meant for exudation of water through a pore.

Hydroponics: A method of growing plants without soil, where nutrients are delivered through water to the plant roots.

Hygroscopic Water: The water which is present around the soil particles in the form of thin vapour layer and held tightly by soil particles of colloidal complex due to adhesive force. This water is also non available water and cannot be drawn/absorbed by the plants.

Hypertonic- Having an osmotic pressure higher than that of an isotonic solution.

Hypocotyl: That part of the embryonic axis which is between the cotyledons and the radicle.

Intracellularly: Occurring within a cell or cells Biogeochemical cycle: The flow of chemical elements and compounds between living organisms and the physical environment.

Isomerization - A reaction that does not change the atomic make-up of a molecule, but rather changes its geometric conformation, yielding a slightly different molecule.

Jasmonic Acid (JA): A plant hormone involved in regulating responses to stress, including herbivore damage and pathogen attack, as well as regulating reproductive processes.

Leghaemoglobin: A haemoglobin like red pigment in the root nodules of leguminous plants is essential for nitrogen fixation.

Light reactions (light dependent reactions): Chemical reactions requiring electromagnetic energy (light) that occur in the thylakoid membrane of the chloroplast to convert light energy into chemical forms ATP and NAPDH.

Lignin: A complex polymer found in the cell walls of plants, providing structural support and resistance to pathogens.

Lumen: Region within the thylakoid membrane where water is split to obtain oxygen. The oxygen diffuses out of the cell, while the protons remain inside to build positive electrical charge inside the thylakoid.

Maturation: The process by which a plant or its parts (such as seeds or fruit) reach full developmental potential, typically involving physiological changes.

Meristem: Tip of a plant's growth.

Mesophyll: The inner tissue of a leaf, consisting of palisade and spongy cells, where photosynthesis occurs.

Metabolism - All the reactions occurring in an organism that participate in the acquisition or conversion of energy for use in the organism.

Micropyle: A small opening in the seed coat through which water enters during seed germination and where the pollen tube enters during fertilization.

Molybdenum: An essential trace element in plant metabolism.

NADPH (Nicotinamide adenine dinucleotide phosphate): It leads to the reduction of nitrate into ammonia for plant assimilation in nitrogen cycle.

Nastic Movement: A non-directional plant movement that occurs in response to environmental stimuli, such as the opening and closing of flowers.

Nicotinamide Adenine Dinucleotide - A coenzyme that participates in oxidation and reduction reactions. An important electron carrier in oxidative phosphorylation.

Nitrate reductases: Molybdoenzymes that reduce nitrate (NO_3^-) to nitrite (NO_2^-). Diffusion Movement of gaseous molecules from higher concentration region to lower concentration region

Nitrification: Oxidation of an ammonium compound into a nitrite or nitrite into nitrate by the action of nitrifying bacteria

Nutrient Uptake: The process by which plants absorb water and essential minerals (such as nitrogen, phosphorus, and potassium) from the soil.

Nutrient: An Element Fundamental to Plant Life.

Osmosis- A process by which molecules of a solvent tend to pass through a semipermeable membrane from a less concentrated solution into more concentrated one.

Oxidation: A reaction that involves the overall loss of electrons from a specific molecule or atom. Can occur with the addition of an oxygen or by the removal of a hydrogen.

Oxidative Phosphorylation - A process occurring in the mitochondria that results in the formation of ATP from the flow of electrons to oxygen.

Oxygen: Tasteless, colourless element, necessary in soil to sustain plant life

Perennial: A plant which completes its life cycle over several to several hundred years.

Phloem: A tissue in vascular plants responsible for the transport of nutrients, particularly sugars produced by photosynthesis, from leaves to other parts of the plant.

Photoperiod: The relationship between the length of light and dark in a 24 hour period

Photoperiodism: A plant's response to the length of day and night, which influences flowering and other developmental processes.

Photosynthesis: A process in which plants convert sunlight into energy sources that can be used inside the cell to sustain life.

Photosystem – A cluster of chlorophyll and other molecules in a thylakoid that harvest the energy of light for photosynthesis

Phototropism: The growth response of plants to light, where stems and leaves grow toward the light source (positive phototropism).

Phytohormone: Plant hormone that control or regulate the germination, growth, metabolism or other physiological activities

Pigment - Colored molecule. A pigment absorbs specific wavelengths of light. Chlorophyll absorbs blue and red light and reflects green light, so it appears green.

Plant Growth: It is the irreversible, quantitative increase in size, mass, and/or volume of a plant or its parts.

Plasmodesmata: are microscopic channels which traverse the cell walls of plant cells and some algal cells, enabling transport and communication between them.

Plasmodesmata: Channels between plant cells that allow for the exchange of materials and communication between cells.

Pollination: The transfer of pollen from the male reproductive organs (anthers) to the female reproductive organs (stigma) of a flower, facilitating fertilization.

Protoplasm: The semi fluid, translucent substance that forms the living matter in all plant

Pyridoxal phosphate: A major coenzyme which is involved in amino acid metabolism.

Reducing sugar: Reducing sugar is a sugar that is capable of acting as a reducing agent because it has a free aldehyde group or a free ketone group. All monosaccharides are reducing sugars, along with some disaccharides, oligosaccharides, and polysaccharides.

Reduction: A reaction that results in the overall gain of electrons to a specific molecule or atom. Can occur with the addition of a hydrogen atom or by the removal of an oxygen atom.

Reproductive growth (or phase): Phase which involves production of flowers, fruits and seeds.

Respiration (Cellular): The process by which plants break down glucose to release energy (ATP), occurring in the mitochondria, and involving oxygen consumption and carbon dioxide production.

Rizobium: Nitrogen-fixing bacteria that form symbiotic relationships with leguminous plants, converting atmospheric nitrogen into forms usable by the plant.

Root Cap: A protective structure at the tip of a plant root that helps the root navigate through the soil and protects the root apical meristem.

Root Hair: Extensions of root epidermal cells that increase surface area for water and nutrient absorption from the soil.

Seed: A structure containing a tiny embryo plant and stored food which can give rise to a new plant under suitable conditions.

Sink: Part or organ of plant to which metabolites and solutes are transferred from source.

Siroheme: A heme like prosthetic group used by some enzyme to attain the six electron reduction of sulfur and nitrogen.

Starch: The principle reserve food materials of the green plants, frequently found in clolourless plastid.

Stomata: Small pores on the surface of leaves and stems that regulate gas exchange (CO_2 in, O out) and water vapor loss in plants.

Symplast: The interconnected network of plant cell cytoplasm linked by plasmodesmata, through which water and solutes can pass.

TCA (Tri Carboxylic Acid) cycle: It is a series of enzyme-catalyzed chemical reactions that form a key part of aerobic respiration in cells.

Tenoplast-Also called vacuolar membrane. Loss of water in form of water drops through hydathodes.

Tissue: Group or layer of similarly specialized cells that together perform certain special functions.

Transaminase (aminotransferase): An enzyme that catalyzes a type of reaction between an amino acid and an α-keto acid.

Translocation: Transport of solutes from site of synthesis or storage to other part of plant where they are required for metabolic activities.

Transpiration:The process by which water is absorbed by roots, transported through the plant, and evaporates from stomata in leaves, contributing to water and nutrient transport.

Urease: An enzyme that changes urea into ammonium carbonate, occurring in microbes *i.e.* bacteria, fungi *etc.*

Vacuole: A fluid filled (cell sap) space bounded by a membrane (tonoplast) found in the cell.

Vegetative growth (or phase): The period of growth between germination and flowering is known as the vegetative phase of plant development.

Water vapour- Un- combined water as moisture in soil.

Wilting coefficient: The level of soil moisture at which water becomes unavailable to plants and caused permanent wilting ensures.

Xylem: Vascular tissue in plants responsible for the transport of water and minerals from the roots to the leaves. It also provides structural support.

Zygote: A fertilized egg cell, formed by the fusion of male and female gametes, that develops into the embryo of a plant.

α- **ketoglutarate:** As anion, α-ketoglutarate (α-KG, also called oxo-glutarate) is an important biological compound.

References

Achard P, Genschik P (2009). "Releasing the brakes of plant growth: how GAs shutdown DELLA proteins". *Journal of Experimental Botany.* 60 (4): 1085-92.

Agrama, H.A.S., Zakaria, A.G., Said, F.B. and Tuinstra, M.R. (1999). Identication of quantitative trait loci for nitrogen use efciency in maize. *Mol. Breed.* 5, 87–195.

An DG, Su JY, Liu QY, Zhu YG, Tong YP, Li JM, Jing RL, Li B, Li ZS (2006). Mapping QTLs for nitrogen uptake in relation to the early growth of wheat (*Triticum aestivum* L.). *Plant Soil* 284: 73–84.

Austin RB (1999). Yield of wheat in the United Kingdom: recent advances and prospects. *Crop Sci* 39: 1604–1610.

Barbottin, A., Lecomte, C., Bouchard, C., Jeuffroy, M.-H., 2005. Nitrogen remobilization during grain filling in wheat: genotypic and environmental effects. *Crop Sci.* 45, 1141–1150.

Barraclough PB, Howarth JR, Jones J, Lopez-Bellido R, Parmar S, Shepherd CE, Hawkesford MJ (2010). Nitrogen efficiency of wheat: genotypic and environmental variation and prospects for improvement. *Eur J Agron* 33: 1–11.

Bernard SM. And Habash DZ. 2009. The importance of cytosolic glutamine synthetase in nitrogen assimilation and recycling. *New Phytol* 192: 608-620.

Bertin, P. and Gallais, A. (2001). Physiological and genetic basis of nitrogen use efciency in maize. II. QTL detection and coincidences. *Maydica* 46: 53–68.

Bogard M, Allard V, Brancourt-Hulmel M, Heumez E, Machet JM, Jeuffroy MH, Gate P, Martre P, Le Gouis J (2010). Deviation from the grain protein concentrationgrain yield negative relationship is highly correlated to post-anthesis N uptake in winter wheat. *J Exp Bot* 61: 4303–4312.

Borrell, A.K., Garside, A.L., Fukai, S. and Reid, D.J. (1998). Season, nitrogen rate, and plant type affect nitrogen uptake and nitrogen use efciency in rice. Aust. *J. Agric. Res.* 49: 829–843.

Borrell, A.K., Hammer, G.L., 2000. Nitrogen dynamics and the physiological basis of stay-green in sorghum. *Crop Sci.* 40, 1295–1307.

Brancourt-Hulmel M, Doussinaut G, Lecomte C, Berard P, LeBuanec B, Trottet M (2003). Genetics improvement of agronomic traits of winter wheat cultivars released in France from 1946 to 1992. *Crop Sci* 43: 37–45.

Brancourt-Hulmel, M., Heumez, E., Pluchard, P., Béghin, D., Depatureaux, C., Giraud, A., LeGouis, J., 2005. Indirect versus direct selection of winter wheat for low input or high input levels. *Crop Sci.* 45, 1427–1431.

Brisson N, Gate P, Gouache D, Charmet G, Oury FX, Huard F. 2010. Why are wheat yields stagnating in Europe? A comprehensive data analysis for France. *Field Crops Res.* 119: 201–212.

Cai, H., Zhou, Y., Xiao, J., Li, X., Zhang, Q. and Lian, X. (2009). Overexpressed glutamine synthetase gene modies nitrogen metabolism and abiotic stress responses in rice. Plant Cell Rep. 28: 527–537.

Caputo, C., Criado, M.V., Roberts, I.N., Gelso, M.A. and Barneix, A.J. (2009). Regulation of glutamine synthetase 1 and amino acids transport in the phloem of young wheat plants. *Plant Physiol. Biochem.* 47: 335–342.

Christopher, J.T., Manschadi, A.M., Hammer, G.L., Borell, A.K., 2008. Developmental and physiological traits associated high yield and stay-green phenotype in wheat. *Aust. J. Agric. Res.* 59, 354–364.

Clárk RB (1983). Plant genotype differences in the uptake, translocation, accumulation, and use of mineral elements required for plant growth. *Plant Soil* 72: 175–196.

Cormier F, Faure S, Dubreuil P, Heumez E, Beauchêne K, Lafarge S, Praud S, Le Gouis J. 2013. A multi environmental study of recent breeding progress on nitrogen use efficiency in wheat (*Triticum aestivum* L.). *Theor Appl Genet* (2013). 126: 3035– 3048.

Cox, M.C., Qualset, C.O., Rains, D.W., 1986. Genetic variation for nitrogen assimilation and translocation in wheat. 3. Nitrogen translocation in relation to grain yield and protein. *Crop Sci.* 26, 737–740.

Davière JM, Achard P (March 2013). Gibberellin signaling in plants. *Development.* 140 (6): 1147-51.

Devlin R.M. 1972. Plant Physiology. Affiliated East-West Press Private Limited. New Delhi.

Dhaka, T.S.; Singh.L. and Kumar.A. Plant Physiology. Pragati Prakashan

Dhugga, K.S., Waines, J.G., 1989. Analysis of nitrogen accumulation and use in bread and durum wheat. *Crop Sci.* 29, 1232–1239.

Dodig D (2005). A high-density genetic map of hexaploid wheat (Triticum aestivum L.). from the cross Chinese Spring × SQ1 and its use to compare QTLs for grain yield across a range of environments. *Theor Appl Genet* 110: 865–880.

Doerge RW (2002). Mapping and analysis of quantitative trait loci in experimental populations. *Nat Rev Genet* 3: 43–52.

Duca M. 2015. Plant Physiology. Springer International Publishing. Switzerland.

Emerson, R. (1957, January). Dependence of yield of photosynthesis in long-wave red on wavelength and intensity of supplementary light. In Science (Vol. 125, No. 3251, pp. 746-746). 1200 NEW YORK AVE, NW, WASHINGTON, DC 20005: *AMER ASSOC ADVANCEMENT SCIENCE.*

Eugene, R. (1969). Govindjee. *Photosynthesis.*

Fischer R, Edmeades G (2010). Breeding and cereal yield progress. *Crop Sci* 50: 85–98.

Fischer, R.A., Wall, P.C., 1976. Wheat breeding in Mexico and yield increases. *Journal of the Australian Institute of Agricultural Science* 42, 139–148.

Fontaine J-X, Ravel C, Pageau K, Heumez E, Dubois F, Hirel B, Le Gouis J (2009). A quantitative genetic study for elucidating the contribution of glutamine synthetase, glutamate dehydrogenase and other nitrogen-related physiological traits to the agronomic performance of common wheat. *Theor Appl Genet* 119: 645–662.

Foulkes M, Hawkesford M, Barraclough P, Holdsworth M, Kerr S, Kightley S, Shewry P (2009). Identifying traits to improve the nitrogen economy of wheat: recent advances and future prospects. *Field Crops Res* 114: 329–342.

Foulkes, M.J., Sylvester-Bradley, R., Scott, R.K., 1998. Evidence for differences between winter wheat cultivars in acquisition of soil mineral nitrogen and uptake and utilization of applied fertiliser nitrogen. *J. Agric. Sci.* (Camb.). 130, 29–44.

Foulkes, M.J., Sylvester-Bradley, R., Weightman, R., Snape, J.W., 2007. Identifying physiological traits associated with improved drought resistance in winter wheat. *Field Crop Res.* 103, 11–24.

Gajua O, V. Allardb, P. Martreb, J.W. Snaped, E. Heumeze, J. LeGouisb, D. Moreaub, M. Bogardb, S. Griffithsd, S. Orfordd, S. Hubbarta, M.J. Foulkesa, 2011. Identification of traits to improve the nitrogen-use efficiency of wheat genotypes. *Field Crops Research* 123 (2011). 139–152.

Gallardo F, Fu J, Canton FR, Garcia-Gutierrez A, Canovas FM, Kirby EG.1999. Expression of a conifer glutamine synthetase gene in transgenic poplar. *Planta* 210: 19–26.

Galloway JN, Townsend AR, Erisman JW, Bekunda M, Cai Z, Freney JR, Martinelli LA, Seitzinger SP, Sutton MA (2008). Transformation of the nitrogen cycle: recent trends, questions, and potential solutions. *Science* 320 (5878): 889–892.

Graybosch R, Peterson C (2012). Specific adaptation and genetic progress for grain yield in Great Plains hard winter wheats from 1987 to 2010. *Crop Sci* 52: 631- 643.

Green A, Berger G, Griffey C, Pitman R, Thomason W, Balota M, Ahmed A (2012). Genetic yield improvement in soft red winter wheat in the eastern United States from 1919 to 2009. *Crop Sci* 52: 2097–2108.

Guarda G, Padovan S, Delogu G (2004). Grain yield, nitrogen-use efficiency and baking quality of old and modern Italian bread wheat cultivars grown at different nitrogen levels. *Eur J Agron* 21: 181–192.

Guo Y, Kong FM, Xu YF, Zhao Y, Liang X, Wang YY, An DG, Li SS (2012). QTL mapping for seedling traits in wheat grown under varying concentrations of N, P and K nutrients. *Theor Appl Genet* 124: 851–865.

Habash DZ, Bernard S, Schondelmaier J, Weyen J, Quarrie SA (2007). The genetics of nitrogen use in hexaploid wheat: N utilisation, development and yield. *Theor Appl Genet* 114: 403–419.

Habash DZ, Massiah AJ, Rong HL, Wallsgrove RM, Leigh RA.2001. The role of cytosolic glutamine synthetase in wheat. *Annals of Applied Biology* 138: 83–89.

Harrison J, Brugiere N, Phillipson B, Ferrario-Mery S, Becker T, Limami A, Hirel B.2000. Manipulating the pathway of ammonia assimilation through genetic engineering and breeding: consequences to plant physiology and plant development. *Plant and Soil* 221: 81–93.

Hedden P (2003). The genes of the green revolution. Trends Genet 19: 5–9. Hirel B, Lea PJ. 2001. Ammonium assimilation. In: Lea PJ, Morot- Gaudry JF, eds. *Plant nitrogen*. Berlin: Springer-Verlag, 79–99.

Heldt, H. W., and Piechulla, B. (2021). Plant biochemistry. Academic Press.

Hirel B, Le-Gouis J, Ney B, Gallais A (2007). The challenge of improving nitrogen use efficiency in crop plants: towards a more central role for genetic variability and quantitative genetics within integrated approaches. *J Exp Bot.* 58 (9): 2369–2387.

Hirel, B., Bertin, P., Quillere´, I. *et al.* (2001). Towards a better understanding of the genetic and physiological basis for nitrogen use efciency in maize. *Plant Physiol.* 125: 1258–1270.

HOPKINS, W.G., HÜNER, N.P.A. 2009: Introduction to Plant Physiology, 4th Edition. John Wiley and Sons,Inc., Hoboken, USA.

http://www.esalq.usp.br/lepse/imgs/conteudo_thumb/Discovery-of-Photosynthesis.pdf

http://www.soilless.org/hydroponics/soilless-farming-hydroponic-crop-cultivation

https://byjus.com/biology/hydroponics/

https://sensorex.com/2019/10/29/hydroponic-systems-explained

https://smartgardenguide.com/types-of-hydroponics-systems

https://www.ledflowergrowlights.eu/illuminate.html

https://www.life.illinois.edu/govindjee/photosynBook/Chapter13.pdf

Hudson, G.S., Evans, J.R., von Caemmerer, S., Arvidsson, Y.B.C., Andrews, T.J., 1992. Reduction of ribulose-1,5-bisphosphate carboxylase oxygenase content by antisense RNA reduces photosynthesis in transgenic tobacco plants. *Plant Physiol.* 98, 294–302.

Ishiyama, K., Inoue, E., Watanabe-Takahashi, A., Obara, M., Yamaya, T. and Takahashi, H. (2004). Kinetic properties and ammonium-dependent regu- lation of cytosolic isoenzymes of glutamine synthetase in Arabidopsis. *J. Biol. Chem.* 279: 16598– 16605.

Jain and Jain: Fundamentals of Biochemistry: S Chand Publication

Jain, V.K. Fundamentals of Plant Physiology, S.Chand and Company pvt. Ltd., New Delhi (revised edition 2011).

Justes, E., Mary, B., Meynard, J.M., Machet, J.M., Thelier-Huche, L., 1994. Determination of a critical nitrogen dilution curve for winter wheat crops. *Ann. Bot.* 74: 397–407.

Kibite S, Evans LE (1984). Cause of negative correlations between grain yield and grain protein concentration in common wheat. *Euphytica* 33: 801–810.

Kjær, B., Jensen, J. and Giese, H. (1995). Quantitative trait loci for heading date and straw characters in barley. *Genome* 38: 1098–1104.

Krapp, A., Saliba-Colombani, V. and Daniel-Vedele, F. (2005). Analysis of C and N metabolisms and of C/N interactions using quantitative genetics. *Photosynth. Res.* 83, 251–263.

Laperche A, Brancourt-Hulmel M, Heumez E, Gardet O, Hanocq E, Devienne-Barret F, Le Gouis J (2007). Using genotype × nitrogen interaction variables to evaluate the QTL involved in wheat tolerance to nitrogen constraints. *Theor Appl Genet* 115: 399–415.

Laperche A, Devienne-Barret F, Maury O, Le Gouis J, Ney B (2006). A simplified conceptual model of carbon/nitrogen functioning for QTL analysis of winter wheat adaptation to nitrogen deficiency. *Theor Appl Genet* 113: 1131–1146.

Lauerer, M., Saftic, D., Quick, W.P., Labate, C., Fichtner, K., Schulze, E.D., Rodermel, S.R., Bogorad, L., Stitt, M., 1993. Decreased ribulose-1,5-bisphosphate carboxylase-oxygenase in transgenic tobacco transformed with antisense RBCS. VI. Effect on photosynthesis in plants grown at different irradiance. *Planta* 190, 332–345.

Lawlor, D.W., 2002. Carbon and nitrogen assimilation in relation to yield: mechanisms are the key to understanding production systems. *J. Exp. Bot.* 53, 773–787.

Le Gouis J, Béghin D, Heumez E, Pluchard P (2000). Genetic differences for nitrogen uptake and nitrogen utilisation efficiencies in winter wheat. *Eur J Agron* 12: 163– 173.

Lea PJ, Miflin BJ 1974 An alternative route for nitrogen assimilation in higher plants. *Nature* (Lond.). 251: 614-616.

Lea, P.J. and Miin, B.J. (2003). Glutamate synthase and the synthesis of glutamate in plants. *Plant Physiol. Biochem.* 41: 555–564.

Lechner E, Achard P, Vansiri A, Potuschak T, Genschik P (December 2006). "F-box proteins everywhere". *Current Opinion in Plant Biology.* 9 (6): 631-8.

Lehninger and Cox: Biochemistry

Lehninger, A. L., Nelson, D. L., Cox, M. M., and Cox, M. M. (2005). Lehninger principles of biochemistry. Macmillan.

Li ZX, Ni ZF, Peng HR, Liu ZY, Nie XL, Xu SB, Liu G, Sun QX (2007b). Molecular mapping of QTLs for root response to phosphorus deficiency at seedling stage in wheat (Triticum aestivum L.). *Prog Nat Sci* 17: 1177–1184.

Lian, X., Xing, Y., Yan, H., Xu, C., Li, X. and Zhang, Q. (2005). QTLs for low nitrogen tolerance at seedling stage identied using a recombinant inbred line population derived from an elite rice hybrid. *Theor. Appl. Genet.* 112: 85–96.

Lima, L., Seabra, A., Melo, P., Cullimore, J. and Carvalho, H. (2006). Post- translational regulation of cytosolic glutamine synthetase of Medicago truncatula. *J. Exp. Bot.* 57: 2751–2761.

Long, S.P., Zhu, X.-G., Naidu, S.L., Ort, D.R., 2006. Can improvement in photosynthesis increase crop yields? *Plant Cell Environ.* 29, 315–330.

Lopez MS, Reynolds MP, Manes Y, Singh RP, Crossa J, Braun HJ (2012). Genetic yield gains and changes in associated traits of CIMMYT spring bread wheat in a "historic" set representing 30 years of breeding. *Crop Sci* 52: 1123–1131.

Loudet, O., Chaillou, S., Merigout, P., Talbotec, J. and Daniel-Vedele, F. (2003). Quantitative trait loci analysis of nitrogen use efciency in Arabidopsis. *Plant Physiol.* 131: 345–358.

Mae T. 1997. Physiological nitrogen efficiency in rice: nitrogen utilization, photosynthesis, and yield potential. In: Ando T, ed. Plant Nutrition for Sustainable Food Production and Environment. Dordrecht, The Netherlands: Kluwer Academic Publishers, 51–60.

Malagoli P, Laine P, Rossato L, Ourry A. 2005. Dynamics of nitrogen uptake and mobilization in field-grown winter oilseed rape (Brassica napus). from stem extension to harvest. *Annals of Botany* 95, 853–861.

Marín-de la Rosa N, Sotillo B, Miskolczi P, Gibbs DJ, Vicente J, Carbonero P, OñateSánchez L, Holdsworth MJ, Bhalerao R, Alabadí D, Blázquez MA (October 2014). "Large-scale identification of gibberellin-related transcription factors defines group VII ETHYLENE RESPONSE FACTORS as functional DELLA partners". *Plant Physiology.* 166 (2):102-23

Masclaux C, Quillere´ I, Gallais A, Hirel B. 2001. The challenge of remobilization in plant nitrogen economy. A survey of physioagronomic and molecular approaches. *Annals of Applied Biology* 138, 69–81.

McGinnis KM, Thomas SG, Soule JD, Strader LC, Zale JM, Sun TP, Steber CM (May 2003). "The Arabidopsis SLEEPY1 gene encodes a putative F-box subunit of an SCF E3 ubiquitin ligase". *The Plant Cell.* 15 (5): 1120-30.

Mengel, K, and Kirkby, E.A. (2001). Principles of Plant Nutrition, 5th ed. Kluwer Academic Publishers, Dordrecht, Netherlands

Mi, C., Liu, J., Chen, F., Zhang, F., Cui, L., Liu, X., 2003. Nitrogen uptake and remobilization in maize hybrids differing in leaf senescence. *J. Plant Nutr.* 26, 447–459.

Miflin BJ, Habash DZ. 2002. The role of glutamine synthetase and glutamate dehydrogenase in nitrogen assimilation and possibilities for improvement in the nitrogen utilization of crops. *Journal of Experimental Botany* 53: 979–987.

Miflin BJ, Lea PJ 1980 Ammonia assimilation. In (BJ Miflin ed). "The Biochemistry of Plants" Vol 5, Academic Press, New York, pp 169 - 202.

Moll RH, Kamprath EJ, Jackson WA (1982). Analysis and interpretation of factors which contribute to efficiency of nitrogen utilization. *Agron. J.* 74: 562-564.

Mortimer ND, Elsayed MA, Horne RE, 2004. Energy and Greenhouse Gas Emissions for Bioethanol Production from Wheat Grain and Sugar Beet. Sheffield, UK: Sheffield Hallam University: Final Report no. 23/1.

Mukherji S., and Ghosh A.K. 2015. Plant Physiology. New Central Book Agency (P) Ltd. New Delhi

Murase K, Hirano Y, Sun TP, Hakoshima T (November 2008). "Gibberellin-induced DELLA recognition by the gibberellin receptor GID1". *Nature.* 456 (7221): 459– 63. doi:10.1038/nature07519. PMID 19037309.

Muurinen S, Slafer GA, Peltonen Sainio P (2006). Breeding effects on nitrogen use efficiency of spring cereals under northern conditions. *Crop Sci* 46: 561–568.

Nelson D.L. and Cox M.M. 2008. Lehninger Principles of Biochemistry. Fifth edition. W.H. Freeman and Company. New York.

Nishesh Sharma. A Textbook of Biochemistry (A conceptual approach to biomolecules). Vigyan bodh Prakashan, Agra.ISBN: 978-93-85763-63-2.

Obara, M., Kajiura, M., Fukuta, Y., Yano, M., Hayashi, M., Yamaya, T. and Sato, T. (2001). Mapping of QTLs associated with cytosolic glutamine synthetase and NADH-glutamate synthase in rice (Oryza sativa L.). *J. Exp. Bot.* 52: 1209–1217.

Ochs, G., Schock, G., Trischler, M., Kosemund, K. and Wild, A. (1999). Com- plexity and expression of the glutamine synthetase multigene family in the amphidiploid crop Brassica napus. *Plant Mol. Biol.* 39: 395–405.

Opik H. and Rolfe S. 2005. The Physiology of Flowering Plants. Cambridge University Press. UK.

Ortiz-Monasterio I, Pena RJ, Sayre KD, Rajaram S (1997b). CIMMYT's genetic progress in wheat grain quality under four N rates. *Crop Sci* 37: 892–898.

Ortiz-Monasterio I, Sayre KD, Rajaram S, McMahon M (1997a). Genetic progress in wheat yield and nitrogen use efficiency under four N rates. Crop Sci 37: 898–904.

Oury, F.X., Be´rard, P., Brancourt-Hulmel, M.et al. (2003). Yield and grain protein concentration in bread wheat: a review and a study of multi-annual data from a French breeding program. *J. Genet. Breed.*57, 59–68.

Palta JA, Fillery IRP. 1995. N application increases pre-anthesis contribution of dry matter to grain yield in wheat grown on a duplex soil. *Australian Journal of Agricultural Research* 46, 507–518.

Pandey S.N. and Sinha B.K. 2015. Plant Physiology. 4th Edition. Vikas Publishing House Pvt. Ltd. New Delhi.

Pandey, B.P., Botany for Degree Students, S.Chand and Company pvt. Ltd., New Delhi

Papakosta, D.K., Garianas, A.A., 1991. Nitrogen and dry matter accumulation, remobilisation, and losses for Mediterranean wheat during grain filling. *Agron. J.* 83, 864–870.

Parry, M.A.J., Madgwick, P.J., Carvalho, J.F.C., Andralojc, P.J., 2007. Prospects for increasing photosynthesis by overcoming the limitations of Rubisco. *J. Agric. Sci.* 145, 31–43.

Potel, F., Valadier, M.H., Ferrario-Me´ry, S. *et al.* (2009). Assimilation of excess ammonium into amino acids and nitrogen translocation in Arabidopsis thaliana– roles of glutamate synthases and carbamoylphosphate synthetase in leaves. *FEBS J.* 276: 4061–4076.

Principles of Biochemistry David L. Nelson and Michael M.Cox (6th edition)

Przulj, N., Momcilovic, V., 2001. Genetic variation for dry matter and nitrogen accumulation and translocation in two-rowed spring barley. II. Nitrogen translocation. *Eur. J. Agron.* 15, 255–265.

Quarrie SA, Steed A, Calestani C, Semikhodskii A, Lebreton C, Chinoy C, Steele N, Pljevljakusic D, Waterman E, Weyen J, Schondelmaier J, Habash DZ, Farmer P, Saker L, Clarkson DT, Abugalieva A, Yessimbekova M, Turuspekov Y, Abugalieva S, Tuberosa R, Sanguineti MC, Hollington PA, Aragues R, Royo A,. Quraishi UM, Abrouk M, Murat F, Pont C, Foucrier S, Desmaizieres G, Confolent C, Rivière N, Charmet G, Paux E, Murigneux A, Guerreiro L, Lafarge S, Le Gouis J, Feuillet C, Salse J. 2011. Cross-genome map based dissection of a nitrogen use efficiency ortho-metaQTL in bread wheat unravels concerted cereal genome evolution. *Plant Journal* 65: 745-756.

Rauh, L., Basten, C. and Buckler, S. (2002). Quantitative trait loci analysis of growth response to varying nitrogen sources in Arabidopsis thaliana. *Theor. Appl. Genet.* 104: 743–750.

Raun, W.R., Solie, J.B., Johnson, G.V., Stone, M.L., Mullen, R.W., Freeman, K.W., Thomason,W.E., Lukina, E.V., 2002. Improving nitrogen-use efficiency in cereal grain production with optical sensing and variable rate application. *Agron. J.* 94, 351–815.

Rengel Z, Marschner P (2005). Nutrient availability and management in the rhizosphere: Exploiting genotypic differences. *New Phytol* 168: 305–312.

Rossato L, Laine´ P, Ourry A. 2001. Nitrogen storage and remobilization in Brassica napus L. during the growth cycle: nitrogen fluxes within the plant and changes in soluble protein patterns. *Journal of Experimental Botany* 52, 1655–1663.

S.K. Verma and Mohit Verma, A textbook of Plant Physiology, Biochemistry and Biotechnology, S.Chand and Company pvt. Ltd., New Delhi.

Saal B, von Korff M, Leon J, Pillen K (2011). Advanced-backcross QTL analysis in spring barley: IV. Localization of QTL × nitrogen interaction effects for yieldrelated traits. *Euphytica* 177: 223–239.

Salisbury, F.B. and Ross C.W. 2005. Plant Physiology. CBS Publishers. New Delhi.

Satyanarayan. A Textbook of Biochemistry

Shanahan, J.F., Kitchen, N.R., Raun, W.R., Schepers, J.S., 2008. Responsive in-season nitrogen management for cereals. Comput. *Electron. Agric.* 61, 51–62.

Shewry PR (2004). Improving the protein content and quality of temperate cereals: wheat, barley and rye. In Cakmak I, Welch R (eds). Impacts of agriculture on human health and nutrition. USDA, ARS, U.S. Plant, Soil and Nutrition Laboratory, Cornell University, USA.

Shimada A, Ueguchi-Tanaka M, Nakatsu T, Nakajima M, Naoe Y, Ohmiya H, Kato H, Matsuoka M (November 2008). "Structural basis for gibberellin recognition by its receptorGID1". *Nature.* 456 (7221): 520-3.

Silva, S.A., de Carvalho, F.I.F., Caetano, V.R., Oliveira, A.C., Coimbra, J.L.M., Vasconcellos, N.J.S., Lorencetti, C., 2000. Genetic basis of stay-green trait in bread wheat. *J. New Seeds* 2, 55–68.

Simmonds, N.W. (1995). The relation between yield and protein in cereal grain. *J. Sci. Food Agric.* 67, 309–315.

STRASBURGER, E. *et al.* 2008: Lehrbuch der Botanik für Hochschulen. 36. Auflage, G. Fischer Verlag,Stuttgart-Jena-New York.

Su JY, Zheng Q, Li HW, Li B, Jing RL, Tong YP, Li ZS (2009). Detection of QTLs for phosphorus use efficiency in relation to agronomic performance of wheat grown under phosphorus sufficient and limited conditions. *Plant Sci* 176: 824–836.

Su, J. Y., Xiao, Y., Li, M., Liu, Q., Li, B., Tong, Y., Jia, J., and Li, Z. (2006). Mapping QTLs for phosphorus-deficiency tolerance at wheat seedling stage. *Plant Soil* 281, 25– 36.

Tabuchi, M., Abiko, T. and Yamaya, T. (2007). Assimilation of ammonium ions and reutilization of nitrogen in rice (*Oryza sativa* L.). *J. Exp. Bot.* 58: 2319– 2327.

Tahir, I.S.A., Nakata, N., 2005. Remobilisation of nitrogen and carbohydrate from stems of bread wheat in response to heat stress during grain filling. *J. Agron. Crop Sci.* 191, 106–115.

Taiz L, Zeiger E. Plant Physiology. 2010. Fifth edition. Sinauer Associates Inc. Massachusetts. USA.

TAIZ, L. and ZEIGER, E. 2002 and 2010: Plant Physiology, 3rd and 5th Edition. The Benjamin Cummings Publishing Company, Redwood City - California.

Tempest DW, Meers JL, Brown CM 1970 Synthesis of glutamate in Aerobacter aerogenes by a hitherto unknown route. *Biochem. J.* 117: 405-407.

Thomas, H., Smart, C.M., 1993. Crops that stay green. Ann. Appl. Biol. 123, 193–219.

Triboï, E., Martre, P., Girousse, C., Ravel, C., Triboï-Blondel, A.M., 2006. Unravelling environmental and genetic relationships between grain yield and nitrogen concentration for wheat. *Eur. J. Agron.* 25, 108–118.

Ueguchi-Tanaka M, Nakajima M, Katoh E, Ohmiya H, Asano K, Saji S, Hongyu X, Ashikari M, Kitano H, Yamaguchi I, Matsuoka M (July 2007). "Molecular interactions of a soluble gibberellin receptor, GID1, with a rice DELLA protein, SLR1, and gibberellin". The Plant Cell. 19 (7): 2140-55. doi: 10.1105/tpc.106.043729. PMC 1955699. PMID 17644730.

Uemura, K., Anwaruzzaman, S., Yokota, A., 1997. Ribulose-1, 5-bisphosphate carboxylase/oxygenase from thermophilic red algae with strong specificity for CO_2 fixation. Biochem. *Biophys. Res. Commun.* 233, 568–571.

Valadier, M.H., Yoshida, A., Grandjean, O., Morin, H., Kronenberger, J., Boutet, S., Raballand, A., Hase, T., Yoneyama, T. and Suzuki, A. (2008). Implication of the glutamine synthetase/glutamate synthase pathway in conditioning the amino acid metabolism in bundle sheath and mesophyll cells of maize leaves. *FEBS J.* 275: 3193–3206.

Van Ginkel, M., Ortiz-Monasterio, I., Trethowan, R., Hernandez, E., 2001. Methodology for selecting segregating populations for improved nitrogen use efficiency in bread wheat. *Euphytica* 119, 223–230.

Van Sanford, D.A., Mackown, C.T., 1987. Cultivar differences in nitrogen remobilisation during grain filling in soft red winter wheat. *Crop Sci.* 27, 295– 300.

Verma, V., A text book of Plant Physiology, Emkay Publications, Delhi

Verma, V., Foulkes, M.J., Caligari, P., Sylvester-Bradley, R., Snape, J., 2004. Mapping quantitative trait loci for flag leaf senescence as a yield determinant in winter wheat under optimal and drought-stressed environments. *Euphytica* 135, 255– 263.

Voet, D. and Voet, J., Biochemistry, Wiley and Sons, 2nd edition

Wang RF, An DG, Hu CS, Li LH, Zhang YM, Jia YG, Tong YP (2011). Relationship between nitrogen uptake and use efficiency of winter wheat grown in the North China Plain. *Crop Pasture Sci* 62: 504–514.

Xu Y, Wang R, Tong Y, Zhao H, Xie Q, Liu D, Zhang A, Li B, Xu H, An D. 2014. Mapping QTLs for yield and nitrogen related traits in wheat: influence of nitrogen and phosphorus fertilization on QTL expression. *Theor Appl Genet.* DOI 10.1007/ s00122-013-2201-y

Zheng Y, Gao Z, Zhu Z (October 2016). "DELLA-PIF Modules: Old Dogs Learn New Tricks". *Trends in Plant Science.* 21 (10): 813–815.

Index

A

B

C

D

N

O

P

R

S